T0276058

SpringerBriefs in Pharmacology and Toxicology

More information about this series at http://www.springer.com/series/10423

Shabir H. Lone · Khursheed Ahmad Bhat
Mohammad Akbar Khuroo

Chemical and Pharmacological Perspective of *Artemisia amygdalina*

Shabir H. Lone
Indian Institute of Integrative Medicine (CSIR-India)
Jammu, Jammu and Kashmir
India

Khursheed Ahmad Bhat
Indian Institute of Integrative Medicine (CSIR-India)
Jammu, Jammu and Kashmir
India

Mohammad Akbar Khuroo
Department of Chemistry
University of Kashmir
Hazratbal, Srinagar, Jammu and Kashmir
India

ISSN 2193-4762 ISSN 2193-4770 (electronic)
SpringerBriefs in Pharmacology and Toxicology
ISBN 978-3-319-25215-5 ISBN 978-3-319-25217-9 (eBook)
DOI 10.1007/978-3-319-25217-9

Library of Congress Control Number: 2015951347

Springer Cham Heidelberg New York Dordrecht London

Printed on acid-free paper

Springer International Publishing AG Switzerland is part of Springer Science+Business Media (www.springer.com)

Contents

Abbreviations

ALP	Alkaline phosphatase
APCI	Atmospheric chemical ionization
$FeCl_3$	Ferric chloride
HPLC	High performance liquid chromatography
LDL	Low density lipoproteins
LOD	Limit of detection
LOQ	Limit of quantification
NAA	Napthelene acetic acid
NCI	National Cancer Institute
OGTT	Oral glucose tolerance test
RI	Retention index
RSD	Relative standard deviation
RT	Retention time
SGOT	Serum glutamate oxaloacetate transaminase
SGPT	Serum glutamate pyruvate transaminase
SL	Sesquiterpene lactones
SRB	Sulphorhodamine B
TCA	Trichloroacetic acid
TIC	Total ion chromatogram

Chapter 1
Introduction

Abstract This chapter deals with the general introduction to the genus *Artemisia*. It highlights the importance of genus *Artemisia*, with a particular focus on the chemistry and biology of the genus. It emphasizes the status of the species *Artemisia amygdalina* which is a critically endangered and endemic plant species of the high altitude Kashmir Himalayas. The morphology along with the conservation status of the plant has been duly addressed.

Keywords Artemisia amygdalina · Endangered species · Veer-Tethwen · Conservation measures · Tissue culture · Micropropagation

1.1 Introduction

The genus *Artemisia* of the family Asteraceae is of great therapeutic and economic importance. *Artemisia* species are widely used medicinal plants in folk/conventional medicine to cure painful menstruation and miscarriage (Lee et al. 1998), inflammations and colds (Hu et al. 1996), arterial hypertension (Zeggwagh et al. 2008; Tahraoui et al. 2007), fevers and malaria (Wright 2002) cancer (Kim et al. 2002), etc. *Artemisia moorcraftiana* is used as a folk medicine for the treatment of malaria in northern areas of Pakistan (Ashraf et al. 2010). *Artemisia annua* listed in Chinese pharmacopeia has been used as a remedy for various fevers including malaria (Mueller et al. 2000); *Artemisia absinthium* as an antispasmodic, febrifuge for restoration of declining mental function/memory enhancer, and inflammation of liver (Wake et al. 2000); *Artemisia afra* for the treatment of malaria, coughs, colds, and headaches (Thring and Weitz 2006); *Artemisia asiatica* in traditional oriental medicine for the treatment of cancer, inflammation, and infections (Bora and Sharma 2010); *Artemisia douglasiana* to treat premenstrual syndrome

S.H. Lone et al., *Chemical and Pharmacological Perspective of* Artemisia amygdalina, SpringerBriefs in Pharmacology and Toxicology,
DOI 10.1007/978-3-319-25217-9_1

and dysmenorrheae (Garcia and Adams 2005); *Artemisia vestita* for inflammatory diseases (Ye et al. 2008); *Artemisia vulgaris* as an anti-inflammatory, analgesic, and in liver diseases, etc. (Gilani et al. 2005). Some species such as *Artemisia annua*, *Artemisia absinthum*, and *Artemisia vulgaris* have been incorporated into the pharmacopeias of several European and Asian countries (Proksch 1992). The traditional Chinese medicine 'Herba Artemisiae Scopariae' is the dried sprout of *Artemisia scoparia* Waldst. et Kit. It can clear away heat, promote diuresis, normalize the function of the gallbladder, and cure jaundice (Committee of National Pharmacopoeia 2005). Herb *Artemisia scoparia* has been frequently used as an important ingredient in many traditional prescriptions (Zhang et al. 2002). It has other pharmacological actions, such as protecting liver, lowering the blood pressure, eliminating fever, sedation, and anti-inflammation, antibacterial, antipathogenic microbes, and antitumor action. Many *Artemisia* species are reported to possess antidiabetic effects and have been used in many countries of Middle East and Turkey as a herbal medicine for the treatment of diabetes, high blood pressure, and gastrointestinal ailments (Al-Shamaony et al. 1994; Subramoniam et al. 1996).

The genus *Artemisia* is known to contain many bioactive compounds which include terpenoids, flavonoids, coumarins, and sterols (Bora and Sharma 2011). Biologically active sesquiterpene lactones with great structural diversity have been isolated from various species of *Artemisia*, eudesmanolides and guaianolides being the most common (Tan et al. 1998; Laid et al. 2008; Thomas et al. 2008; Ruiukar et al. 2011). Sesquiterpene lactones (SL) are among the prominent natural products found in *Artemisia* species and are largely responsible for the importance of these plants in medicine and pharmacy. During the past three decades, sesquiterpene lactones have emerged as one of the largest group of plant products with over 3000 naturally occurring substances. Research on sesquiterpene lactones began within a few years of the NCI program. More than fifteen hundred publications have reported SL's anticancer and anti-inflammatory properties. The SLs are almost exclusively derived from the family Asteraceae to which the genus *Artemisia* belongs. Extracts rich in SLs have gained considerable interest for treating human diseases such as inflammation, headache, and infections. The SL-derived drugs from thapsigargin, artemisinin, and parthenolide are now in cancer clinical trials (Lu 2002; Zhang 2008; Berger 2005; Singh and Verma 2002). Artemisinin, which is majorly isolated from *Artemisia annua* is not only the current drug of choice for malaria, but also possesses profound cytotoxicity against tumor cells (Efferth 2007); Arglabin from *Artemisia myriantha* was employed for treating certain types of cancer in the former USSR (Wong and Brown 2002).

Recent phytochemical, pharmacological, and clinical studies on various species have confirmed their ethanomedicinal properties. An exhaustive literature survey revealed that the genus *Artemisia* has a vast range of biological activities mainly antimalarial, cytotoxic, anti-inflammatory, anti-hepatotoxic, antibacterial, antifungal, and antioxidant (Bora and Sharma 2011).

To this genus belongs a plant species *Artemisia amygdalina* Decne, locally called Veer-Tethween. It is an endemic medicinal plant of the Kashmir valley, belonging to the family Asteraceae and grows in the subalpine region of Kashmir Himalaya and North-West Frontier Province of Pakistan (Dar et al. 2006). The plant extract is used locally for the treatment of epilepsy, piles, nervous disorders, cough, cold, fever, and pain (Rasool et al. 2012). The women folk of the valley use it for treating amenorrhea and dysmenorrhea (Qaiser 2006). As a consequence of overharvest and deforestation, this plant is considered as the critically endangered and endemic species of Kashmir valley (Dar et al. 2006). Based on the occurrence of this plant in a unique and particular eco-geographical region, i.e., Kashmir, and being an endangered plant species of genus *Artemisia*, it becomes imperative to look for the chemical and biological potential of this valuable medicinal plant. To this effect chemical and biological exploitation of the plant has been carried out, which will be discussed in the following chapters. For facilitating germplasm conservation efforts of *Artemisia amygdalina,* Khan et al. (2014) has developed an in vitro propagation of this plant.

1.2 Morphology

Artemisia amygdalina Decne (Fig. 1.1) belongs to genus *Artemisia*. It is an erect, up to 1.5 m tall perennial herb. Stems many from the base, shallowly to deeply grooved, glabrous, younger shoots hairy. Leaves are almost sessile, simple, lamina narrowly elliptic-lanceolate, $9 - 15(-16) \times 1 - 3.5$ cm, undivided, serrate, teeth incurved, gland-tipped, hoary tomentose beneath, glabrous green above, gradually attenuate and setose-auricled at the base, and apex long acuminate. Capitula numerous, heterogamous, ±pendulous, c. 3–4 mm across, peduncles 1–1.5 mm long, in $20 - 35 \times 5 - 6$ cm panicle with suberect lateral branches up to 5×1 cm. Involucre 3-4-seriate, outermost phyllaries narrowly ovate, slightly hairy outside, c. 3×1.25 mm, ciliate on membranous margins, acute; innermost elliptic-oblong, 3.5-4 × c. 1.5 mm, glabrous, margins broadly membranous. Receptacle glabrous, ±convex, c. 1 mm in diameter. Florets up to 25, all fertile;

Fig. 1.1 *Artemisia amygdalina* Decne (Photography taken on July 29, 2011 in Gurez Valley)

marginal florets female, 8–10 with 2-toothed, c. 1 mm long corolla, style branches flat; disk-florets bisexual, 12–15, with 5-toothed urceolate, c. 1.5 mm long, pale, glabrous, basally constricted corolla, anther appendages ±obtuse, exserted. Cypselas terete or cylindrical, c. 1 mm long, smooth.

The flowering period of the plant is generally July–September. This species can be easily differentiated from all others in the region by its simple, serrate leaves, which are hoary tomentose beneath and glabrous green above, appearing like Salix leaves from a distance. The closest allies of *A. amygdalina* Decne. are Chinese endemics, *Artemisia anomala* S. Moore and *Artemisia viridissima* Pamp. The evidence of the close relationship of these two species is found chiefly in the leaf and achene morphology. *Artemisia anomala* S. Moore differs from it in shorter and elliptic leaves, whereas *Artemisia viridissima* Pamp. has leaves which closely resemble those of *Artemisia amygdalina* Decne. in shape and size (Rasool et al. 2013).

1.3 Conservation Measures

For facilitating germplasm conservation efforts of *Artemisia amygdalina,* Khan et al. (2014) developed an in vitro propagation of this plant. This method was developed by use of different plant growth regulators in various concentrations (Table 1.1). Nodal explants of *Artemisia amygdalina* were incubated on full strength MS medium, supplemented with BAP (Cytokinins) in combination with NAA (auxins). Complete callus development was observed in 2-week-old cultures on all tested media. It was found that more number of shoots developed on nodal explants exposed to 10 μM NAA and 10 μM BAP with an average of 36 shoots per nodal plant than other treatments after 42 days (Table 1.1). The treatment with BAP alone did not induce shoot organogenesis from node explants of *Artemisia amygdalina;* however, addition of NAA to BAP containing medium also induced shoot organogenesis. Regenerated node explants were separated and sub-cultured on MS medium supplemented with various concentrations of BAP and Kinetin for further shoot proliferation and growth with half strength MS medium giving the best shoot proliferation results after 30 days of culture. Root initials were found to be present after 2 weeks of culture with the highest rate of root development observed on half strength MS medium after 42 days (Table 1.2). The highest percentage on the rooting response has been observed for half strength MS medium with about 7 ± 0.3 roots per regenerated node with an average length of 8.2 cm. However, addition of 2,4-D caused increased root formation with indirect rooting. Hence, use of 2,4-D proved to help in root formation of callus. The rooted plantlets survived ex-vitro transplantation in the greenhouse normal conditions without any supplemental light. Plants fertilized with half strength MS solution every 24 h were able to grow and develop well-formed leaves with characteristic morphology within 3 months.

Table 1.1 Effect of different concentrations of BAP and NAA on the shoot regeneration from nodal explants after 6 weeks

Induction media	Callusing	Regeneration	No of shoots/ node	Average size of shoots (cm)	% Age response
MS + BAP (0.88 μM) + NAA (0.54 μM)	No response	–	–	–	–
MS + BAP (5 μM) + NAA (5 μM)	No response	–	–	–	–
MS + BAP (7.5 μM) + NAA (7.5 μM)	Moderate callusing	Direct regeneration	18 ± 0.1	4.4	80
MS + BAP (7.5 μM) + NAA (10 μM)	Low callusing	Direct regeneration	21	4.2	70
MS + BAP (10 μM) + NAA (7.5 μM)	Low callusing	Direct regeneration	24	4.7	90
MS + BAP (10 μM) + NAA (10 μM)	Low callusing	Direct regeneration	36	4.8	90
MS + BAP (10 μM) + NAA (12.5 μM)	Moderate callusing	–	–	–	60
MS + BAP (12.5 μM) + NAA (10 μM)	High callusing	–	–	–	50
MS + BAP (12.5 μM) + NAA (12.5 μM)	High callusing	–	–	–	60
MS + BAP (15 μM) + NAA (15 μM)	No response	–	–	–	–

Table 1.2 Effect of plant growth regulators on the rooting of invitro raised shoots after 6 weeks

Rooting media	No of roots		Mean size of roots		% Age response
	Regenerated shoot tip	Regenerated node	Regenerated shoot tip (cm)	Regenerated node (cm)	
MS	3 ± 0.1	5 ± 0.2	6.3	7.1	90
Half strength MS	6 ± 0.3	7 ± 0.3	7.5	8.2	100
MS + 2,4-D (1 μM)	9 ± 0.6	9 ± 0.2	3.7	3.9	80
MS + 2,4-D (1 μM)	5 ± 0.2	6 ± 0.3	3.2	3.4	80

1.4 Conclusion

In conclusion, a safe and efficient protocol for the conservation of a rare but medicinally important *Artemisia amygdalina* has been developed. The developed protocol may serve as an effective measure for the conservation of other medicinally important but rare and endangered plant species.

References

Al-Shamaony L, Al-Khazraji MS, Twaij HA (1994) J Ethnopharmacol 43:167–171
Ashraf M, Hayat MQ, Jabeen S, Shaheen N, Khan MA, Yasmin GJ (2010) Med Plant Res 4:112–119
Berger TG (2005) Oncol Rep 14:1599–1603
Bora KS, Sharma A (2010) A Review 3:325–328
Bora KS, Sharma A (2011) Pharm Biol 49:101–109
Committee of National Pharmacopoeia (2005) Pharmacopoeia of People's Republic of China, vol 1. Press of Chemical Industry, Beijing, p 166
Dar AR, Dar GH, Reshi Z (2006) Endangered Species Update 23:34–39
Efferth T (2007) Planta Med 73:299–309
Garcia C, Adams J (2005) Healing with medicinal plants of the west-cultural and scientific basis for their use. Abedus Press, La Crescenta
Gilani AH, Yaeesh S, Jamal Q, Ghayur MN (2005) Phytother Res 19:170–172
Hu J, Zhu Q, Bai S, Jia Z (1996) Planta Med 62:477–478
Khan M, Ganie BA, Kamili AN, Khalid GM (2014) Int Res J Pharm 5:115–118
Kim JH, Kim HK, Jeon SB, Son KH, Kim EH, Kang SK, Sung ND, Kwon BM (2002) Tetrahedron Lett 43:6205–6208
Laid ME, Hegazy F, Ahmed AA (2008) Phytochem Lett 1:85–88
Lee SJ, Chung HY, Maier CGA, Wood AR, Dixon RA, Mabry TJ (1998) J Agric Food Chem 46:3325–3329
Lu L (2002) Chin J Integr Trad West Med 22:169–171
Mueller MS, Karhagomba IB, Hirt HM, Wemakor E (2000) J Ethnopharmacol 73:487–493
Proksch P (1992) Artemisia. Springer, Berlin, pp 357–377
Qaisar M (2006) Flora Pakistan 207:120–121
Rasool R, Ganai BA, Kamili AN, Akbar S (2012) Nat Prod Res 26:2103–2106
Rasool R, Ganai BA, Akbar S, Kamili AN (2013) Chin J Nat Med 11:0377–0384
Ruikar AD, Jadhav RB, Phalgune UD, Rojatkar SR, Puranik VG (2011) Helv Chim Acta 94:73–77
Singh NP, Verma KB (2002) Arch Oncol 10:279–280
Subramoniam A, Pushpangadan RS, Rajasekharan S, Evans DA, Latha PG, Valsaraj R (1996) J Ethnopharmacol 1:13–17
Tahraoui A, El-Hilaly J, Israili ZH, Lyoussi B (2007) J Ethnopharmacol 110:105–117
Tan RX, Zheng WF, Tang HQ (1998) Planta Med 64:295–302
Thomas OP, Ortet R, Prado S, Mouray E (2008) Phytochemistry 69:2961–2965
Thring TSA, Weitz FM (2006) J Ethnopharmacol 103:261–275
Wake G, Court J, Pickering A, Lewis R, Wilkins R, Perry E (2000) J Ethnopharmacol 69:105–114
Wong H, Brown GD (2002) Phytochemistry 59:529–536
Wright WC (2002) Artemisia. Taylor and Francis, London, pp 1–344
Ye Y, Fang-Yuan G, Xing-Xin W, Yang S, Yi-Hua L, Ting C (2008) J Ethnopharmacol 120:1–6
Zeggwagh NA, Farid O, Michel JB, Eddouks M, Moulay I (2008) Methods Find Exp Clin Pharmacol 30:375–381
Zhang ZY (2008) Chin J Integr Med 6:134–138
Zhang QW, Zhang YX, Zhang Y, Xiao YQ, Wang ZM (2002) China J Chin Mater Med 27:202–204

Chapter 2
Phytochemical Screening and HPLC Analysis of *Artemisia amygdalina*

Abstract This chapter deals with the qualitative analysis of wild and tissue culture raised regenerants of *Artemisia amygdalina*, for the amount of bioactive principles particularly the antimalarial compound, artemisinin. Phytochemical screening of extracts revealed the presence of terpenes, alkaloids, phenolics, tannins (polyphenolics), cardiac glycosides, and steroids in wild (aerial, inflorescence) and tissue culture regenerants (in vitro grown plant, callus,and greenhouse acclimatized plants). Further, HPLC of *A. amygdalina* extracts has revealed the presence of artemisinin in petroleum ether extracts of wild aerial part, tissue culture raised plant, and greenhouse acclimatized plants. Acetonitrile and water in 70:30 ratios at a flow rate of 1 ml/min have been optimized as mobile phase. It has been observed that wild inflorescences and callus do not produce artemisinin.

Keywords Qualitative analysis · HPLC · Artemisinin · Tissue culture · Regenerants · Wild plants

2.1 Qualitative Analysis

Qualitative analysis has been carried out for both wild and tissue culture obtained plants. Five plant samples of *Artemisia amygdalina*, viz., wild aerial (A), wild inflorescence (I), in vitro cultured plants (T), callus (C),and acclimatized greenhouse plants (G) have been sequentially screened for the presence of bioactive compounds. Different quantitative tests performed by the authors' group include:

S.H. Lone et al., *Chemical and Pharmacological Perspective of* Artemisia amygdalina, SpringerBriefs in Pharmacology and Toxicology,
DOI 10.1007/978-3-319-25217-9_2

2.1.1 Presence of Tannins

Ferric chloride has been used to detect the presence of tannins in the extract solutions of *A. amygdalina*. As a normal procedure, 2 ml of 5 % $FeCl_3$ has been added to 2 ml aqueous extract of each sample. Yellow brown precipitate indicated the presence of tannins (Jigna and Sumitra 2007; Rasool et al. 2010). It has been observed that tannins were present in the methanolic extracts of wild inflorescence (I), in vitro cultured plants (T), callus (C), and acclimatized greenhouse plants (G) (Table 2.1).

2.1.2 Presence of Alkaloids

Dragendroffs test has been used to detect the presence of alkaloids. The methanolic extracts of wild aerial (A) wild inflorescence (I), in vitro cultured plants (T), callus (C) tested positive for the presence of alkaloids while the acclimatized greenhouse plants (G) tested negative for the alkaloid content (Table 2.1).

Table 2.1 Qualitative phytochemical screening of wild and tissue culture raised regenerants of *A. amygdalina*

Bioactive agents	Type of extract	Presence(+)/Absence(–)				
		Wild aerial (A)	Wild infloresсence (I)	Callus (C)	Tissue culture grown plants (T)	Greenhouse raised plants (G)
Alkaloid	Methanol	+	+	+	+	–
Phenolics	Methanol	+	–	–	–	+
Tannins	Methanol	–	+	+	+	–
Cardiac glycosides	Methanol	+	+	–	+	+
Flavonoids	Methanol	–	–	–	–	–
	Aqueous	–	–	–	–	–
Saponins	Methanol	–	–	–	–	–
Terpenes	Methanol	–	–	–	–	–
	Aqueous	–	–	–	–	–
	Pet. ether	+	+	+	+	+
	Ethyl acetate	+	–	–	+	+
Steroids	Methanolic	–	–	–	–	–
	Pet. ether	+	+	+	+	+
Resins	Methanolic	–	–	–	–	–

2.1.3 Presence of Saponins

The presence of Saponins was confirmed by the addition of a few drops of sodium bicarbonate solution to 2 ml aqueous extract of all samples (Harborne 1973; Oguyemi 1979; Trease and Evans 1989; Sofowora 1993; Jigna and Sumitra 2007). As per the authors' group all the methanolic extracts of *A. amygdalina* Decne tested negative for the presence of saponins (Table 2.1).

2.1.4 Presence of Cardiac Glycosides

Keller Kiliani test was used to test the presence of cardiac glycosides (Sofowora 1993; Rasool et al. 2010). All the methanolic extracts of *A. amygdalina* tested positive for the presence of cardiac glycosides except for the callus extracts (Table 2.1).

2.1.5 Presence of Terpenes

All the methanolic and aqueous extracts of *A. amygdalina* have been found to be devoid of terpenes. All the low polar petroleum ether extracts have been found to contain terpenes. In case of ethyl acetate extracts, only the wild aerial (A), tissue culture grown plants (T) and greenhouse acclimatized petroleum ether extracts have been found to contain terpenes (Table 2.1).

2.1.6 Test for Steroids

Leiberman Buchard test was successfully applied for the presence of steroids (CCRUM 1987). All the low polar petroleum ether extracts were found to contain steroids. However, the high polar methanolic extracts were found to be devoid of steroids (Table 2.1).

2.1.7 Test for Flavonoids

Shinoda's test involves addition of few drops of conc. HCl followed by 0.5 g of zinc turnings to about 2 ml aqueous or methanolic extracts and then boiling for a few minutes to furnish magenta red or pink color has been used to detect the presence of flavonoids (Martinez and Valencia 2003; Jigna and Sumitra 2007). Both the methanolic and aqueous extracts of *A. amygdalina* Decne have been found to be negative for the occurrence of flavonoids (Table 2.1).

2.1.8 Presence of Phenolics

Ferric chloride test has been used to detect the presence of phenolics (CCRUM 1987; Martinez and Valencia 2003). Only the methanolic extracts of wild aerial (A) and greenhouse acclimatized plants (G) have been found to contain phenolics (Table 2.1).

2.1.9 Presence of Resins

Methanolic extracts have been treated with 5 ml acetic anhydride. Solutions were heated and subsequently cooled. 0.5 ml of sulfuric acid was added to all sample solutions. Since no color change was found, therefore it was predicted that these extracts do not contain resins (Table 2.1).

2.2 HPLC Analysis

Standardization of artemisinin in *A. amygdalina* by HPLC-UV has been carried out using the optimized conditions. Detection has been done at 210 nm. Acetonitrile and water in 70:30 ratios have been found to be suitable for quantification. 2.5, 5, 10, 15, 20 μl injections of standard and unknown sample have been run. The typical chromatogram of artemisinin with r.t 6.7 is shown in Fig. 2.1. Among various extracts prepared such as methanolic, ethyl acetate, and aqueous extracts only the low polar petroleum ether extracts have shown the presence of

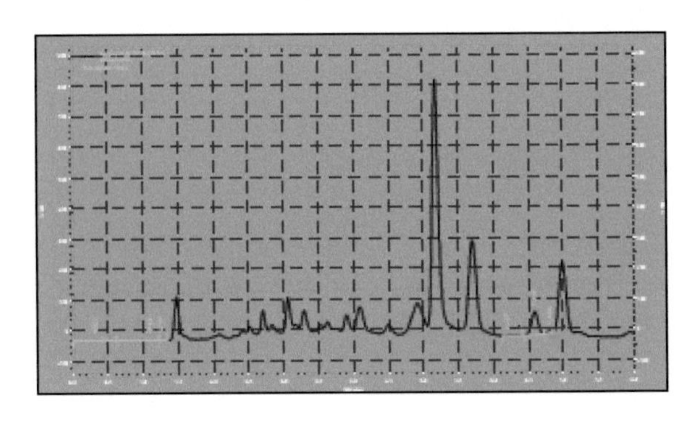

Fig. 2.1 Chromatogram of standard artemisinin with RT 6.7

artemisinin. Petroleum ether extracts of wild aerial (A), in vitro grown (T) and greenhouse acclimatized plants (G) have been found to contain artemisinin while inflorescence (I) and callus (C) have been found to be devoid of this wonder molecule. The peak area ratios of standard and sample solutions have been calculated via external standard method of the Chrom-Quest software. The calibration curve depicted linearity ($y = 6.52951e - 006x + 0.033$). The authors have highlighted their results in Table 2.2 and Figs. 2.1, 2.2, 2.3, 2.4 and 2.5.

The variation in phytochemical profile has been attributed to somaclonal variations that arise during tissue culture cycle and acclimatization process. Appearance of desirable or undesirable variants is a chance event. Genotype, source of explants, duration of culture,and culture conditions all have effect on regenerants.

Table 2.2 Percentage of artemisinin (Mean ± SD % *w/w*) in *A. amygdalina* wild and tissue culture raised regenerants using petroleum ether extracts

Type of extracts	Wild aerial (A)	Wild inflorescence (I)	Tissue culture raised plants (T)	Callus (C)	Greenhouse grown plants (G)
Percentage	0.36 % ± 0.015	0	0.25 % ± 0.07	0	0.28 % ± 0.012
Goodness to fit (r^2)	0.998	0	0.9993	0	0.997

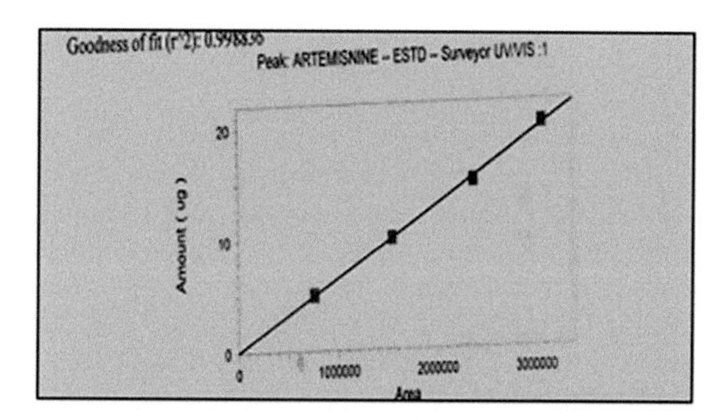

Fig. 2.2 Calibration curve of standard artemisinin

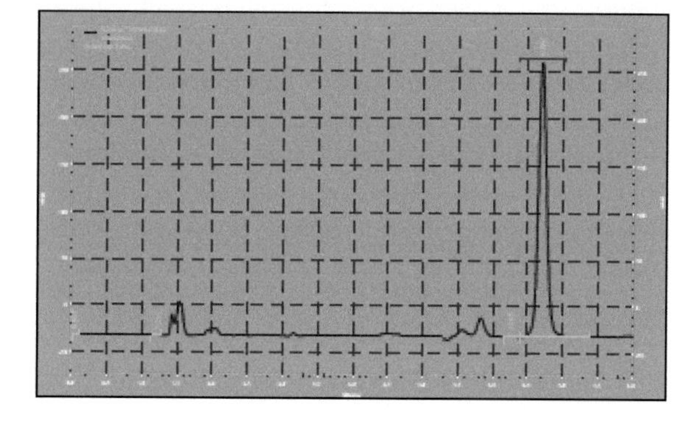

Fig. 2.3 Chromatogram of *A. amygdalina* (wild aerial part)

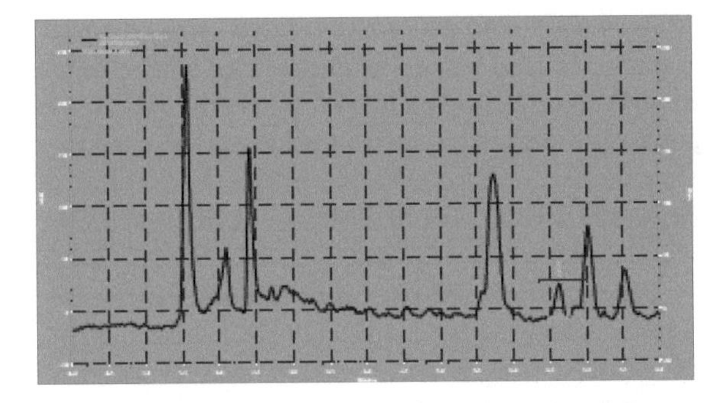

Fig. 2.4 Chromatogram of *A. amygdalina* (in vitro raised)

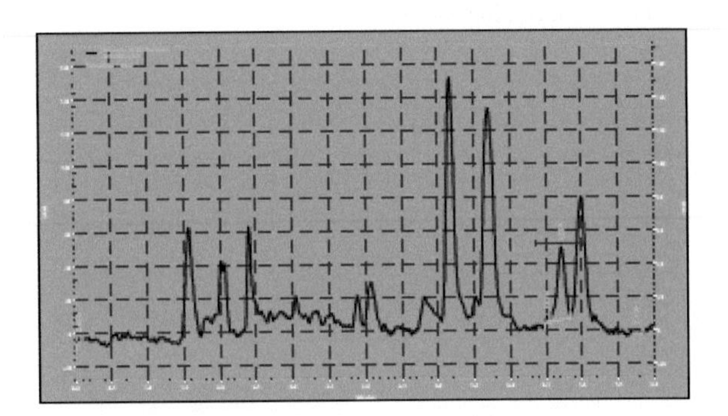

Fig. 2.5 Chromatogram of *A. amygdalina* (Greenhouse raised plants)

Artemisinin, a sesquiterpene lactone has been detected in wild aerial part, invitro grown plant, greenhouse acclimatized plant and was found absent in callus and inflorescences. Artemisinin possesses weak chromophore, so derivatization using NaOH and acetic acid could help in detection (Mannan et al. 2010), but acetic acid used in the procedure here has led to noise production with pressure fluctuations in column. UV absorption of artemisinin in HPLC system, with inbuilt default wavelength range of 190–800 nm, has been found to be high enough that allowed the quantification without alkaline hydrolysis treatment, which is in conformity with the findings of Ferreira and Gonzalez 2009. It also maintained the system in equilibrium. So work has been carried without alkaline hydrolysis treatment.

As per (Singh and Sarin 2010) *Artemisia scoparia* callus culture is an alternative to *Artemisia annua* for the production of artemisinin. The yield of artemisinin in *A. annua* has been reported to be higher in aerial plant parts (0.015 %) in comparison to callus culture (0.001 %), which is in accordance with that of *A. amygdalina,* where aerial parts also showed higher concentration of artemisinin (0.36 %) but not in callus (Table 2.2). As per Mannan et al. 2010, the highest artemisinin concentration has been detected in the leaves (0.44 ± 0.03 %) and flowers (0.42 ± 0.03 %) of *A. annua*, followed by the flowers (0.34 ± 0.02 %) of *Artemisia bushriences* and leaves (0.27 ± 0 %) of *Artemisia dracunculus*. Varying concentrations of artemisinin in various species ranging from 1.38 % in *A. annua* leaves in Switzerland (Delabays et al. 1993), 0.86 % in *A. annua* leaves in Vietnam Wallaart et al. 1999), 0.79 % in *A. annua* in leaves in China (Charles et al. 1990), 0.0006 % in *Artemisia cina* in Indonesia (Aryanti et al. 2001), 0.2 % *Artemisia seibri* in Iran (Arab et al. 2006) have been reported worldwide.

References

Arab HA, Rahbari S, Rassouli A, Moslemi MH, Khosravirad FDA (2006) Trop Anim Health Prod 38:497–503
Aryanti Bintang M, Ermayanti TM, Mariska I (2001) Annales Bogorienses 8:11–16
CCRUM (1987) Physico-chemical standards of unani medicinal formulations. Nice Printing Press, Publication Number, 26
Charles DJ, Simon JE, Wood KV, Heinstein P (1990) J Nat Prod 53:157–160
Delabays N, Collet G, Benakis A (1993) Acta Hort 330:203–207
Ferreira JFS, Gonzalez JM (2009) Phytochem Anal 20:91–97
Harborne JB (1973) Phytochemical methods. Chapman and Hall Ltd., London, pp 49–188
Jigna P, Sumitra VC (2007) Turk J Biol 31:53–58
Mannan A, Ahmed I, Arshad A, Asim FM, Qureshi AR, Hussain I, Mirza B (2010) Malaria J 9:310
Martinez A, Valencia G (2003) Manual de practicas de Farmacognosia y Fitoquimia: 1999. 1. Medellin: Universidad de Antiquia; Marcha fotiquimica, pp 59–65
Oguyemi AO (1979). In: Sofowora A (ed) Proceedings of a conference on African medicinal plants. Ife-Ife: Univ Ife, pp 20–22
Rasool R, Ganai BA, Kamili AN, Akbar S, Masood A (2010) Pak J Pharm Sci 23:399–402
Singh A, Sarin R (2010) Bangladesh J Pharmacol 5:17–20
Sofowora A (1993) Medicinal plants and traditional medicine in Africa. Spectrum Books Ltd., Ibadan, Nigeria, p 289
Trease GE, Evans WC (1989) Pharmacognosy, 11th edn. Brailliar Tiridel Can., MacMillian Publishers, Ibadan
Wallaart T, Pras N, Quax W (1999) Planta Med 65:723–728

Chapter 3
Volatile Composition of *Artemisia amygdalina*

Abstract This chapter deals with the analysis of mono and sesquiterpenoid composition in the leaves and stem of *Artemisia amygdalina* Decne obtained by hydrodistillation using a combination of capillary GC-FID and GC-MS analytical techniques. A total of 25 components in the leaf and 32 components in the stem essential oils have been reported. The leaf essential oil majorly consists of monoterpene hydrocarbons (38.2 %) and oxygenated monoterpenes (43.8 %) together constituting 82.0 % of the total oil composition. The principal components that have been identified include sabinene (10.2 %), p-cymene (14.7 %), 1,8-cineole (17.5 %), and borneol (19.8 %). The stem essential oil comprises monoterpene hydrocarbons (66.1 %), oxygenated monoterpenes (12.8 %) and sesquiterpene hydrocarbons (11.2 %). The major constituents were α-pinene (6.0 %), camphene (10.4 %), β-pinene (40.2 %), and borneol (5.7 %). A comparative analysis of the essential oil obtained from the leaves of micropropagated plants of *A. amygdalina* with that of the leaves of field grown parent plants has also been reported. The oil yield from the micropropagated plants has been found to be (0.05 % v/w) than the oil yield obtained from field-grown plants (0.2 % v/w). The major constituents of the field-grown plants include p-cymene (21.0 %), 1,8-cineole (24.9 %), α-terpineol (5.9 %), β-caryophyllene (4.7 %), germacrene-D (4.0 %), while as the major constituents of the micropropagated plants reported include p-cymene (11.3 %), 1,8-cineole (10.2 %), borneol (7.9 %), α-longipinene (5.5 %), α-copaene (5.5 %) and β-caryophyllene (17 %). The essential oil from field-grown plant is further dominated by the presence of oxygenated monoterpenes (41.5 %), monoterpene hydrocarbons (35.9 %) and sesquiterpene hydrocarbons (16.3 %) while as the essential oil of micropropagated plants is known to be characterized by sesquiterpene hydrocarbons (40.0 %), oxygenated monoterpenes (25.2 %) and monoterpene hydrocarbons (21.6 %).

S.H. Lone et al., *Chemical and Pharmacological Perspective of* Artemisia amygdalina, SpringerBriefs in Pharmacology and Toxicology,
DOI 10.1007/978-3-319-25217-9_3

Keywords Oil analysis · Stem oil · Leaf oil · Monoterpene hydrocarbons · Oxygenated monoterpenes · Sabinene · p-cymene

3.1 Oil Analysis

The volatile components of the leaf and stems of *Artemisia amygdalina* Decne have been reported by Rather et al. (2011, 2012) using GC-FID and GC-MS analysis. Identification of the essential oil constituents has been done on the basis of Retention Index (RI, determined with respect to homologous series of n-alkanes ((C_9–C_{24}), Polyscience Corp., Niles IL) under the same experimental conditions), co-injection with standards (Sigma Aldrich and standard isolates), MS Library search (NIST 98 and WILEY), by comparing with the MS literature data (Jennings and Shibamoto 1980; Adams 2007). The relative percentages of the individual components have been calculated based on GC peak area (FID response) without using correction factors. The different essential oil constituents of the leaf and stem that have been reported in *A. amygdalina* have been shown in Table 3.1, in order of their elution from RTX-5 column. The oil yields calculated on dry weight basis have been found to be 0.1 and 0.05 % (v/w), respectively, for the leaf and stem. Capillary GC-FID and GC-MS analysis has led to the identification of 25 components in the leaf and 32 components in the stem essential oils. The total ion chromatograms (TIC) of the two essential oils are shown in Figs. 3.1 and 3.2. The leaf essential oil is dominated by the presence of monoterpene hydrocarbons (38.2 %) and oxygenated monoterpenes (43.8 %) together constituting 82.0 % of the total oil composition. Sesquiterpene hydrocarbons comprise 10.7 % while oxygenated sesquiterpenes constitute only 4.1 %. The principal components of the leaf oil include sabinene (10.2 %), p-cymene (14.7 %), limonene (3.0 %), 1,8-cineole (17.5 %), borneol (19.8 %), and α-terpineol (2.3 %). The stem essential oil however is dominated by monoterpene hydrocarbons constituting 66.1 % of the total oil composition. Oxygenated monoterpenes constituted 12.8 %, sesquiterpene hydrocarbons accounted for 11.2 % while as oxygenated sesquiterpenes constituted 8.1 % of the total oil composition. The graphical representation of various mono and sesquiterpenes in the two essential oils has been shown in Fig. 3.3. The major constituents of the stem essential oil are reported to be α-pinene (6.0 %), camphene (10.4 %), β-pinene (40.2 %), borneol (5.7 %) and pinocarvyl acetate (3.5 %). The chemical structures of the major compounds present in the two essential oils along with their corresponding mass spectra have been shown in Figs. 3.4, 3.5, 3.6, 3.7, 3.8, 3.9 and 3.10. On comparative analysis of the leaf and stem essential oils of *A. amygdalina*, it is evident that the two essential oils show considerable qualitative similarity but differ quantitatively.

Another study carried out by the same research group reports the comparative analysis of the essential oil of micropropagated and field grown plants of *A. amygdalina*. The average yield of the essential oils obtained from the leaves of micropropagated plants of *A. amygdalina* has been 0.05 % (v/w dry weight). Higher oil yield

Table 3.1 Various constituents in the leaf and stem essential oils of *Artemisia amygdalina*

Compound	RI	Peak area % in leaf	Peak area % in stem
Santolina triene	909	1.0	0.1
α-Thujene	924	–	0.2
A-pinene	936	2.6	6.0
Camphene	950	1.3	10.4
β-pinene	973	10.2	1.8
3-Carene	977	2.3	40.2
p-cymene	1008	–	0.5
Limonene	1024	14.7	3.1
1,8-cineole	1028	3.0	2.6
(Z)-β-ocimene	1031	17.5	2.2
γ-Terpinene	1036	0.9	0.4
α-Terpinoline	1058	2.2	0.6
Camphor	1086	–	0.2
2-Nonene-1-ol	1146	1.6	–
Borneol	1152	–	0.3
Terpenen-4-ol	1169	19.8	5.7
α-Terpinol	1177	0.4	–
Piperitone	1188	2.3	0.5
Pinocarvyl acetate	1252	2.2	–
Myrtenyl acetate	1311	1.5	3.5
α-Longipinene	1324	–	0.6
α-Copane	1353	–	0.1
β-Borbonene	1377	1.0	1.5
β-Caryophyllene	1390	0.4	–
(Z)-β-Farnescene	1419	4.2	2.4
α-Humulene	1443	–	0.3
α-Curcumene	1454	1.3	3.1
Germacrene D	1483	0.6	–
BicycloGermacrene	1487	2.3	2.7
δ-Cadinene	1500	–	0.5
E-Nerolidol	1525	0.9	0.6
Germacrene-D-4-ol	1561	–	0.4
Spathulenol	1574	–	0.6
Caryophyllene oxide	1577	–	0.7
α-cadinol	1586	1.9	1.8
Ledene oxide	1652	–	3.1
α-Bisabolol	1687	2.2	–

of 0.2 % (v/w dry-weight) has been obtained from the leaves of field-grown plants. The two essential oils showed a similar chemical profile but differed in the relative percentages of the major components. 29 components belonging to different

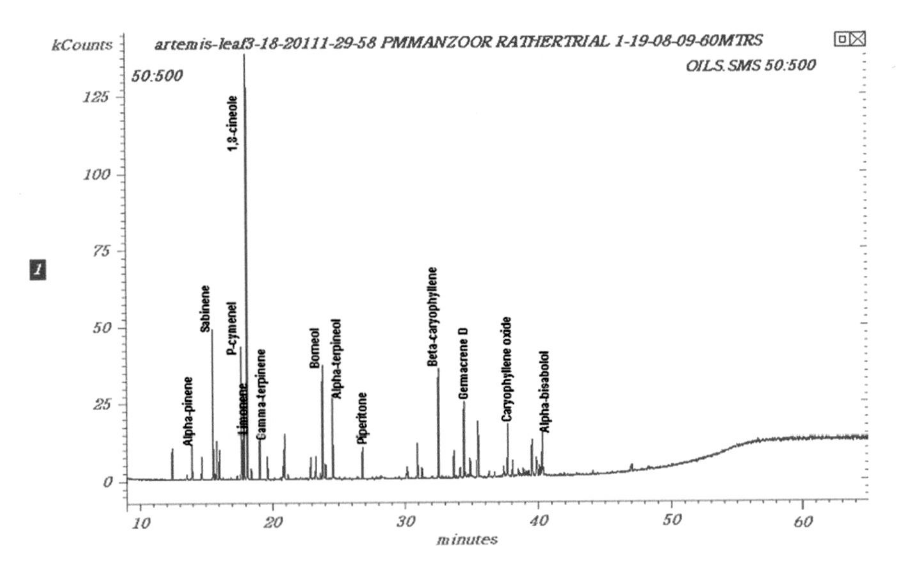

Fig. 3.1 TIC of the leaf essential oil of *Artemisia amygdalina*

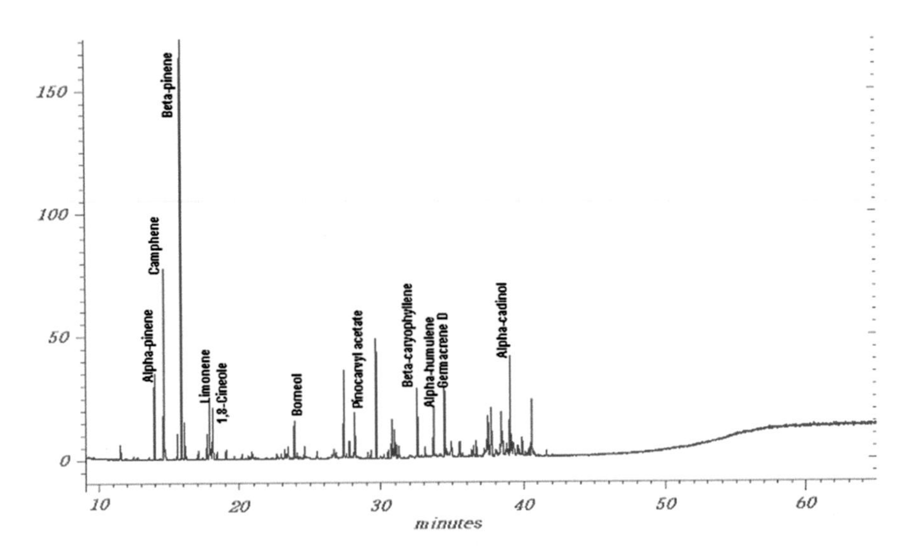

Fig. 3.2 TIC of the stem essential oil of *Artemisia amygdalina*

terpenoid classes have been identified in the essential oil of field-grown plants as compared to twenty-seven components in the essential oil of micropropagated plants. The TICs of the essential oils of field-grown and micropropagated plants have been shown in Figs. 3.11 and 3.12, respectively. The compounds identified in the field-grown and micropropagated plants of *A. amygdalina* along with their retention indices (RI) and relative percentages have been presented in Table 3.2.

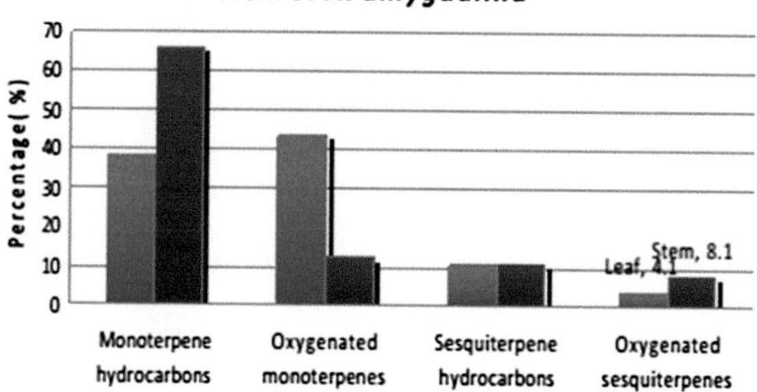

Fig. 3.3 Compound classes in two essential oils

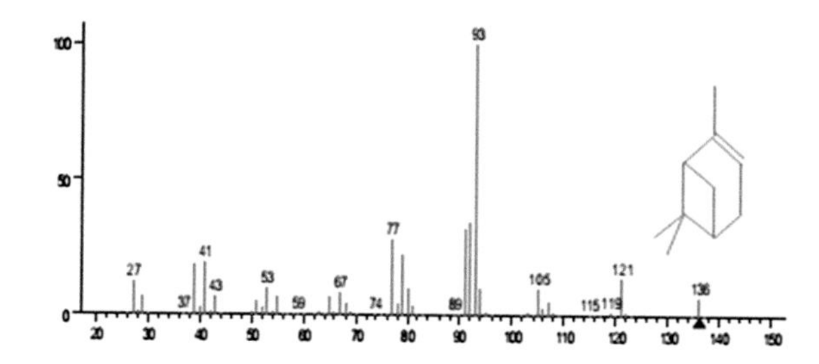

Fig. 3.4 Mass spectrum of α-pinene

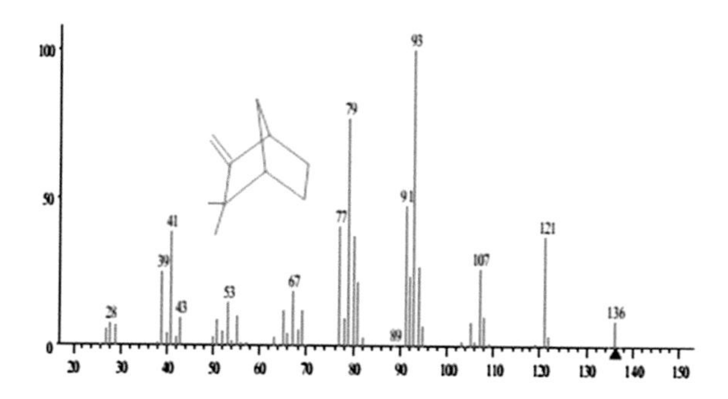

Fig. 3.5 Mass spectrum of camphene

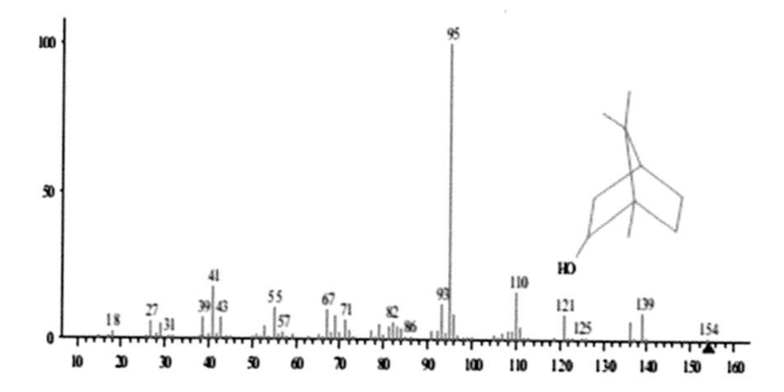

Fig. 3.6 Mass spectrum of borneal

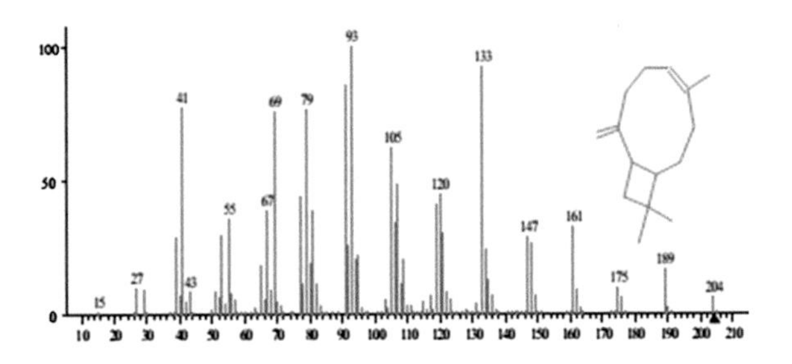

Fig. 3.7 Mass spectrum of p-cymene

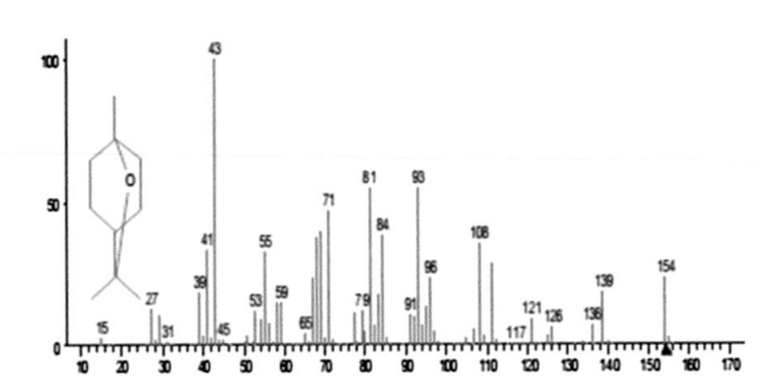

Fig. 3.8 Mass spectrum of β-pinene

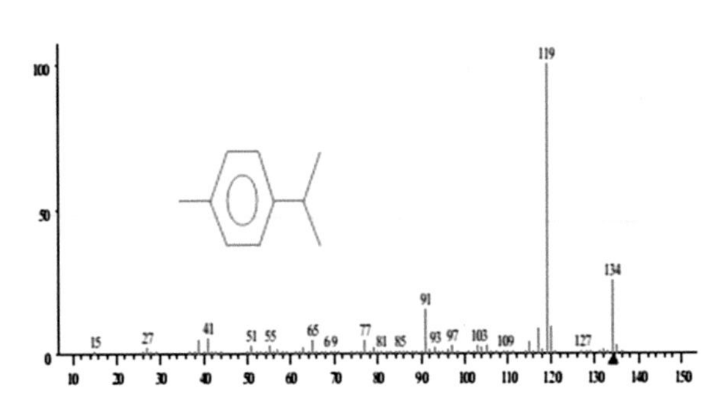

Fig. 3.9 Mass spectrum of β-caryophyllene

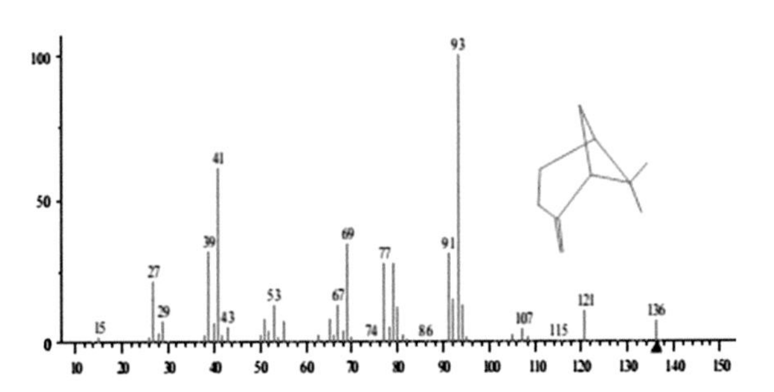

Fig. 3.10 Mass spectrum of 1,8-cineole

The essential oil from the field-grown plants was dominated by the presence of oxygenated monoterpenes (41.5 %), monoterpene hydrocarbons (35.9 %) and sesquiterpene hydrocarbons (16.3 %). The major components of the oil were

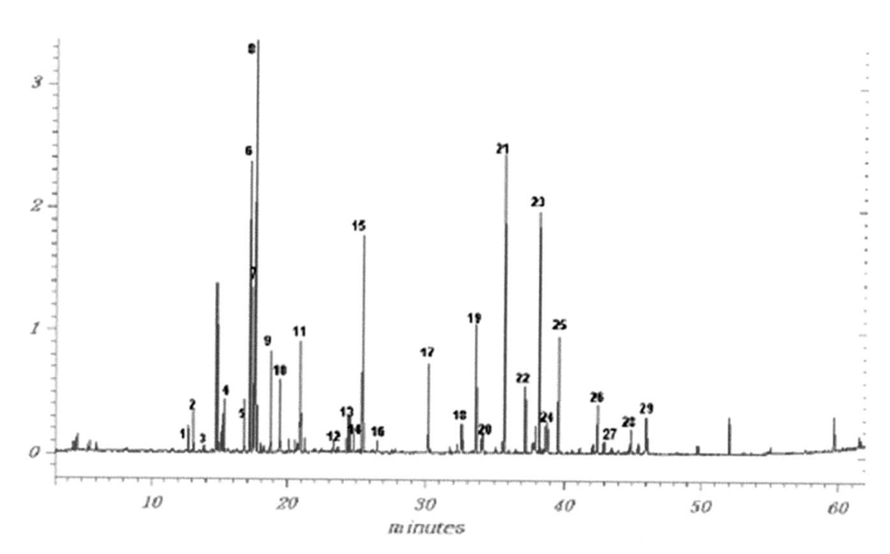

Fig. 3.11 TIC of the field grown plants

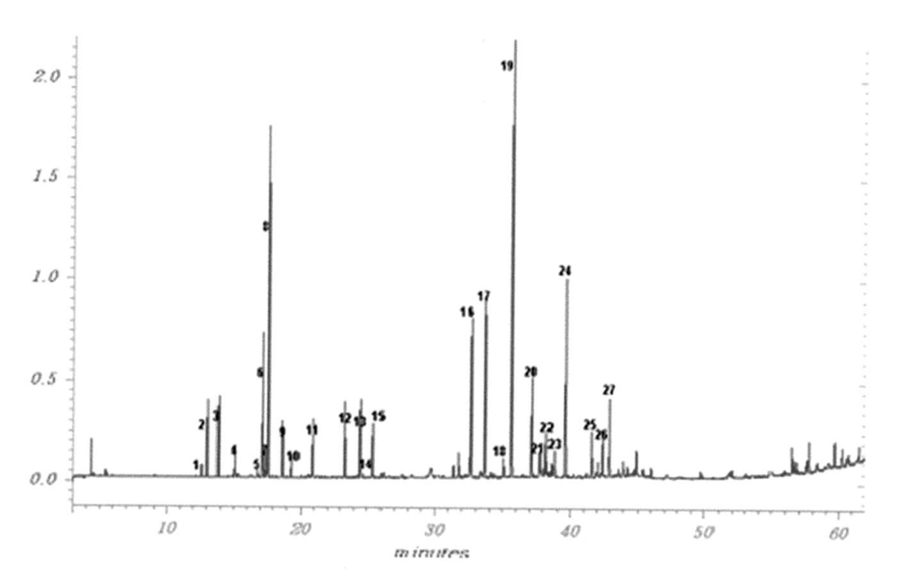

Fig. 3.12 TIC of the essential oil of micropropagated plants

p-cymene (21.0 %), 1,8-cineole (24.9 %), α-terpineol (5.9 %), β-caryophyllene (4.7 %), and germacrene D (4.0 %). On the other hand, the essential oil of micropropagated plants was characterized by sesquiterpene hydrocarbons (40 %), oxygenated monoterpenes (25.2 %) and monoterpene hydrocarbons (21.6 %). The major components were p-cymene (11.3 %), 1,8-cineole (10.2 %), borneol (7.9 %), α-longipinene (5.5 %), α-copaene (5.5 %), and β-caryophyllene (17.0 %). Both the

Table 3.2 Comparative profile of the field grown (FG) and micropropagated plants (MP)

Compound	RI	Peak area % (FG)	Peak area % (MP)
α-Thujene	927	0.8 ± 0.02	0.6 ± 0.01
α-pinene	935	3.9 ± 0.01	3.5 ± 0.03
Camphene	951	0.4 ± 0.01	0.4 ± 0.02
β-pinene	978	1.8 ± 0.04	1.8 ± 0.03
α-Terpinene	1012	0.5 ± 0.01	0.6 ± 0.01
p-cymene	1025	21 ± 0.06	11.3 ± 0.03
Limonene	1029	5.0 ± 0.1	0.9 ± 0.04
1,8-cineole	1032	24.9 ± 0.5	10.2 ± 0.1
γ-Terpinene	1052	2.5 ± 0.02	2.5 ± 0.03
(E)-Sabinene hydrate	1097	1.6 ± 0.05	0.6 ± 0.01
1-Terpineol	1136	2.7 ± 0.04	1.3 ± 0.05
Camphor	1142	0.3 ± 0.02	3.3 ± 0.07
Borneol	1170	3.5 ± 0.06	7.9 ± 0.10
Terpenen-4-ol	1175	0.6 ± 0.03	0.1 ± 0.01
α-Terpineol	1189	5.9 ± 0.10	1.8 ± 0.02
Piperitone	1251	tr	–
β-Dehydroelsholtzia ketone	1307	2.0 ± 0.01	–
α-Longipinene	1356	0.7 ± 0.04	5.5 ± 0.20
α-Copane	1376	2.9 ± 0.02	5.5 ± 0.02
α-Gurjunene	1411	0.1 ± 0.01	0.4 ± 0.02
β-Caryophyllene	1417	4.7 ± 0.10	17.0 ± 0.21
α-Humulene	1455	1.9 ± 0.02	0.8 ± 0.01
Germacrene D	1489	4.0 ± 0.10	4.4 ± 0.10
β-Silenene	1491	0.2 ± 0.01	1.5 ± 0.01
α-muurolene	1507	0.4 ± 0.02	0.7 ± 0.01
δ-Cadinene	1527	1.4 ± 0.01	4.2 ± 0.31
Caryophyllene oxide	1588	0.9 ± 0.02	0.9 ± 0.02
Longiborneol	1596	0.1 ± 0.01	1.2 ± 0.02
α-Bisabolol	1689	0.8 ± 0.02	0.2 ± 0.02

essential oils have shown a low percentage of oxygenated sesquiterpenes indicating a similarity in the biosynthetic pathways of field-grown and micropropagated plants. The graphical representation of the relative percentages of various mono and sesquiterpenes present in the two essential oils is depicted in Fig. 3.13. On comparative analysis of the two essential oils, it is evident that the field-grown plants showed higher content of p-cymene (21.0 %), 1,8-cineole (24.9 %) and α-terpineol (5.9 %), while as in case of the micropropagated plants, these three compounds were present in relatively lesser percentages viz. p-cymene (11.3 %), 1, 8-cineole (10.2 %), and α-terpineol (1.8 %). However, the micropropagated plants contained higher percentage of sesquiterpene hydrocarbons like β-caryophyllene (17.0 %), α-longipinene (5.5 %), α-copaene (5.5 %), and δ-cadinene (4.2 %).

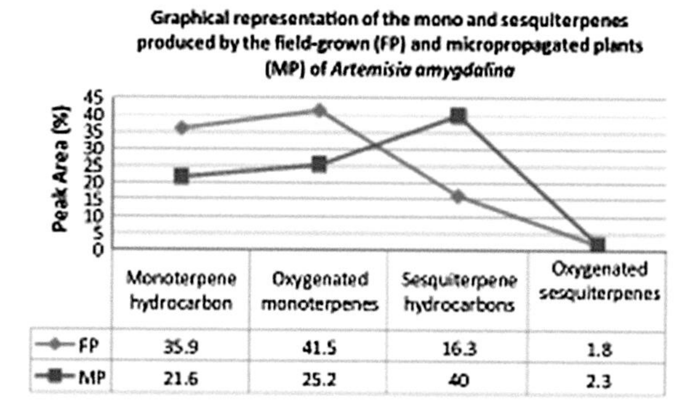

Fig. 3.13 Graphical distribution of the various mono and sesquiterpenes present in the field grown and micropropagated plants

A thorough literature survey has revealed that the essential oil composition of the *Artemisia* genus has been thoroughly investigated and the chemodiversity in the oil composition has led to many oil-dependent chemotypes for the genus, among which 1,8-cineole/camphor chemotype appear to be predominant in most *Artemisia* species (Lutz et al. 2008; Govindraj et al. 2008; Lawrence 1982; Lamberg 1982). In some *Artemisia* species such as *A. annua*, *A. vulgaris*, *A. diffusa*, *A. santonicum*, *A. spicigera*, *A. afra*, *A. abiatica*, *A. austriaca* and *A. pedemontana*, bornane derivatives (camphor, borneol and bornyl acetate) and 1,8-cineole are the major biochemical marker compounds (Perez-Alonso et al. 2003; Kordali et al. 2005).

3.2 Conclusion

In conclusion, the essential oil obtained from the micropropagated plants of *A. amygdalina* is comparable to that obtained from the field-grown parent plants. The two essential oils display qualitative similarity but show quantitative difference in the mono and sesquiterpenoid constituents. All the biochemical marker constituents present in the field-grown plants are also present to greater or lesser extent in the micropropagated plants. The study thus shows that since this plant species is critically endangered and the micropropagated plants of *A. amygdalina* may be a nice alternative to the wild-growing plants.

References

Adams RP (2007) Identification of essential oil components by gas chromatography/mass spectrometry. Allured Publishing Corp, Carol Stream

Govindraj S, Ranjithakumari BP, Cioni PL, Flamini G (2008) J Bios Bioengineer 05:176–183

Jennings W, Shibamoto T (1980) Qualitative analysis of flavor and fragrance volatile by glass capillary gas chromatography. Academic Press Inc, New York

Kordali S, Cakir A, Mavi A, Kilic H, Yildirim A (2005) J Agric Food Chem 53:1408–1416

Lamberg S (1982) Perfum Flavor 7:58–63
Lawrence BM (1982) Perfum Flavor 6:37–38
Lutz DL, Alviano DS, Alviano CS, Kolodziejczyk PP (2008) Phytochemistry 69:1732–1738
Perez-Alonso MJ, Velasco NJ, Paula P, Sanz J (2003) Biochem Syst Ecol 31:77–84
Rather MA, Sofi SN, Dar BA, Ganai BA, Masood A, Qurishi MA, Shawl AS (2011) J Phar Res 4:1637–1639
Rather MA, Ganai BA, Kamili AN, Qayoom M, Akbar M, Rasool R, Wani SH, Qurishi MA, Shawl AS (2012) Acta Physiol Plant 34:885–890

Chapter 4
Phytochemical Analysis and Chemobiological Standardization of *Artemisia amygdalina*

Abstract This chapter deals with the bioactivity guided isolation of *Artemisia amygdalina* Decne. The hexane extracts of both shoot and root parts of *Artemisia amygdalina* Decne have displayed potent cytotoxic effects. Phytochemical analysis of these active extracts has led to the isolation of six cytotoxic constituents, viz., 7,22-ergostadien-3β-ol (**1**), ludartin (**2**), 5-hydroxy-6,7,3′,4′-tetramethoxyflavone (**3**) (from shoots) and *trans*-matricaria ester (**4**), diacetylenic spiroenol ether (**5**) and *cis*-matricaria ester (**6**) (from root) from this plant. The constituents have been identified using spectral and analytical techniques in the light of literature. Sulphorhodamine B cytotoxicity screening of the isolated constituents has been carried out against four human cancer cell lines including lung (A-549), leukemia (THP-1), prostate (PC-3), and colon (HCT-116) cell lines. Ludartin (**2**) exhibited the highest cytotoxicity with IC_{50} values of 7.4, 3.1, 7.5 and 6.9 μM against lung (A-549), leukemia (THP-1), prostate (PC-3), and colon (HCT-116) cancer cell lines, respectively. To test against in vitro skin cancer models [human dermal fibroblasts (CRL-1635)], all the isolates have been further subjected to 3-(4,5-dimethylthiazol-yl)-diphenyl tetrazolium bromide (MTT) cytotoxicity screening. Ludartin being highly cytotoxic has been evaluated against mouse melanoma (B16F10) and human epidermoid carcinoma (A-431) cells by MTT assay displaying IC_{50} values of 6.6 μM and 19.0 μM respectively. Finally a simple and reliable HPLC method has been developed (RP-HPLC-DAD) and validated for the simultaneous quantification of these cytotoxic constituents in *A. amygdalina* Decne. Excellent specificity and high linearity for all the standard calibration curves, having regression coefficients of the respective linear equations in the range of 0.9962 – 0.9999, has been observed. Relative recovery rates varied between 98.37 ± 0.90 and 105.15 ± 1.74 with relative standard deviation less than 4 %. Based on these results, the developed method features good quantification parameters, accuracy, and precision and can serve as effective quality control method for standardization of *A. amygdalina* Decne.

S.H. Lone et al., *Chemical and Pharmacological Perspective of* Artemisia amygdalina, SpringerBriefs in Pharmacology and Toxicology,
DOI 10.1007/978-3-319-25217-9_4

Keywords Isolation · Quantification · Validation · MTT evaluation · SRB evaluation · Ludartin

4.1 Introduction

Artemisia (sagebrush or wormwood) is one of the abundant and widely distributed genus of the family Asteraceae with around 300–400 species of herbs and shrubs known for their chemical constituents that are extensively used in food and pharmaceutical industry (Kwak et al. 1997). *Artemisia*s are popular plants, used for the treatment hepatitis, inflammation cancer, and infections by various microorganisms (Tan et al. 1998; Kim ct al. 2002). An exhaustive literature survey has revealed that the genus *Artemisia* exhibits a vast array of biological activities mainly antimalarial, cytotoxic, antihepatotoxic, antibacterial, antifungal, anti-inflammatory, and antioxidant. Terpenes, flavones, coumarins, and sterols are the main chemical classes present in various *Artemisia* species (Bora et al. 2011). Highly bioactive sesquiterpene lactones with great structural variations have been isolated from various species of *Artemisia* with eudesmanolides and guaianolides being the most common (Laid et al. 2008; Thomas et al. 2008; Ruikar et al. 2011; Efferth 2007; Wong and Brown 2002). This chapter unfolds the secondary metabolite constitution of this plant with regard to isolation, characterisation, cytotoxicity evaluation, and chemobiological standardization of *Artemisia amygdalina*. This chapter therefore envisages contributing towards:

(i) The integrated assessment of the chemical composition and exploitation potential of *A. amygdalina* Decne.
(ii) The biological potential of the isolated chemical constituents from *A. amygdalina* Decne.
(iii) The development and validation of an RP-HPLC method for simultaneous quantification of the chemical constituents of *A. amygdalina* Decne for use in herbal industry.

4.2 Phytochemistry

Preliminary screening of hexane extracts has been carried out using SRB assay against human lung (A-549), colon (HCT-116), prostate (PC-3), and leukemia (THP-1) cell lines. (Fig. 4.1)

Both the hexane extracts of root and shoot showed potent cytotoxicity almost against all the tested cell lines. Thus a bioactivity guided approach has been taken into account and 60.0 g of the hexane extract of shoot has been subjected to column chromatography over silica gel to afford compounds **1** (700 mg), **2** (2.0 g), and **3** (100 mg) using hexane-EtOAc as eluent with increasing polarity of 5, 10 and 25 % EtOAc respectively (Fig. 4.2). Similarly the hexane extract of root (18.0 g) has

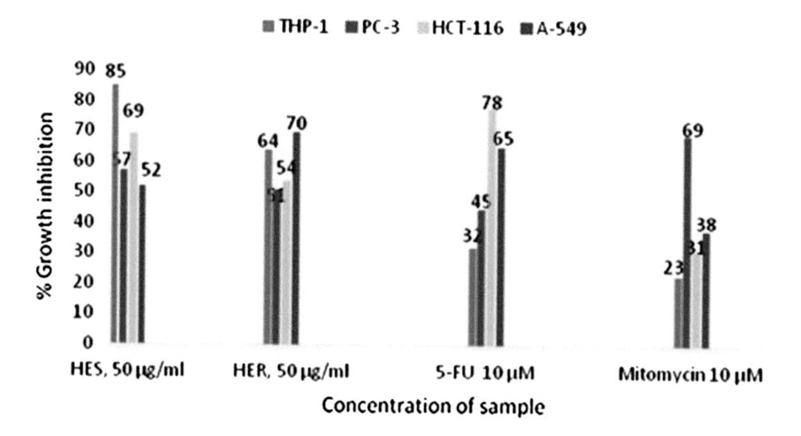

Fig. 4.1 Cytotoxic profile of the hexane extracts of *Artemisia amygdalina* Decne (HES=Hexane extract Shoot) and (HER=Hexane Extract Root)

Fig. 4.2 Structures of the isolated constituents from *Artemisia amygdalina* Decne

been subjected to fractionation using flash chromatography system and afforded fraction I (4 g) and fraction II (10 g) with 5 and 15 % EtOAc-Hexane respectively. Repeated Column chromatography of fraction I, carried on normal-phase silica gel (60–120) utilizing long glass column with a narrow lumen and an isocratic-elution system comprising 1 % EtOAc-Hexane has yielded three compounds: **4** (50 mg), **5** (1.5 g) and **6** (65 mg), while as that of fraction II yielded **1** (200 mg; also isolated from shoot). The structures of the compounds have been depicted in Fig. 4.2 and have been characterized using spectral data in the light of literature.

Compound **1** has been isolated as a white solid powder from both the shoot and root extracts of *Artemisia amygdalina* Decne with APCI mass at *m/z* 398.35 [M^+] corresponding to molecular formula $C_{28}H_{46}O$, having melting point in the range 147–152 °C. The MS of **1** exhibited two important fragments at *m/z* 383.36 (M^+–CH_3) and *m/z* 255.2117 (M^+–H_2O-alkyl side chain). IR showed strong absorption band at 3430 (OH), 1627, and 146 cm^{-1} (double bond). Combined analysis of ^1H and ^{13}C NMR of compound **1** showed the presence of 28 signals assignable to six methyl groups with singlets for C (18) at δ_H 0.742 ppm, δ_C 12.28 ppm and C (19) at δ_C 12.92 ppm, δ_H 0.797 ppm, doublets for C (21) at δ_C 21.68 ppm, δ_H 1.037 ppm, $J = 6.8$ Hz, C (26) at δ_C 21.42 ppm, δ_H 0.871 ppm, $J = 6.4$ Hz, C (27) δ_C 19.23 ppm, δ_H 0.821 ppm, $J = 2.4$ Hz, and C (28) δ_C 19.35 ppm, δ_H 0.839 ppm, $J = 2.4$ Hz), a trisubstituted olefin [δ_H 5.37 (1H, d, $J = 6$ Hz, H-C (6), δ_C 121.73 C (6), 140.78 C (5)], a disubstituted olefin [δ_H 5.174 (1H, dd, $J_1 = J_2 = 8.4$ Hz, H-C(22) and δ_H 5.037 (1H, dd, $J_1 = J_2 = 8.4$ Hz, H-C (23), δ_C 138.32 C (22), and 129.305 C (23)], eight methylenes, three quaternary carbons and a C-3 hydroxyl group (δ_H 3.57, δ_C 71.2 ppm). The above data was in complete agreement with that reported for 7, 22-ergostadien-3β-ol (Smania et al. 1999). Hence **1** has been identified as 7,22-ergostadien-3β-ol.

Compound **2** has been isolated as a white crystalline powder having sodiated molecular ion peak at *m/z* 269.00 $[M + Na]^+$ corresponding to the molecular formula $C_{15}H_{18}O_3$ having melting point of 110 °C. The IR spectrum showed signals at 1760 (γ-lactone), 1451, 1381 (C=C) and 1130 (C-O-C) cm^{-1}. ^{13}C NMR revealed the presence of 15 carbon resonances assignable to two methyls, four methylenes, four methines, and five quaternary carbons. Presence of two proton resonances at δ ppm: 6.12 (1H, d, $J = 3.5$ Hz, H–C (13_a)) and 5.38 (1H, d, $J = 3.5$ Hz, H–C (13_b)) along with the ^{13}C NMR signal at δ ppm: 169.56 and IR stretching frequency (1760 and 1639 cm^{-1}) are strong indicative of an α-methylene-γ-lactone substructure. Further proton resonances at δ ppm: 3.68 [1H, t, $J = 10.8$, 10.8 Hz, H-C (6)], 3.10 [1H, d, $J = 10.8$ Hz, H–C (5)] with high J values are an indicative of the fact that proton on carbon no. 6 possesses a *trans*-diaxial relationship both with carbon 5 and carbon 7 protons (Bhadane and Shafizadeh 1975). Further two peaks at δ ppm: 63.6 (CH) and 67.68 (C) were assignable to an oxirane nucleus. Finally HMBC and HSQC correlations along with previous literature reports (Geissman and Griffith 1972; Sosa et al. 1989) have confirmed **2** as Ludartin.

Compound **3** has been isolated as a yellow powder showing APCI mass at *m/z* 359.11 $[M + 1]^+$ which corresponds to molecular formula $C_{19}H_{18}O_7$. The melting point of the molecule was found to be 179 °C. IR spectrum exhibited bands at

3432 cm^{-1} (OH), 1659 cm^{-1} (C=C conjugated to C=O), and 1516, 1461 cm^{-1} (aromatic region). ^{13}C NMR revealed the presence of 19 signals, 15 of which are assignable to the aromatic region of the whole flavonoid nucleus and rest four signals at δ (ppm): 60.90, 56.36, 56.40, and 56.13 were assigned to four methoxyl groups. The ^{1}H NMR showed two resonance signals of AB type at δ (ppm): 7.53 (1H, d, $J = 8.5$ Hz*)* and 6.98 (1H, d, $J = 8.5$ Hz) assignable to H-C (5´) and H-C (6´) of the B-ring, respectively. Further three other proton resonances at δ (ppm): 7.34 (1H, s), 6.61 (s, 1H), and 6.56 (s, 1H) were assigned to H-C (2´), H-C (8) and H-C (3) protons respectively. HSQC, HMBC and literature data (Harbone et al. 1986) confirmed the structure of **3** as 5-Hydroxy-6,7,3´,4´-tetramethoxyflavone.

Compound **4** has been isolated as a dark brown powder having APCI mass at *m/z* 173.01 $[M + 1]^{+}$ which corresponds to molecular formula $C_{11}H_{8}O_{2}$. IR exhibited resonances at 2140 (C≡C), 1670 (C=C in conjugation with C=O), and 1120 (C-O-C) cm^{-1}. ^{13}C NMR revealed the presence of 11 carbons, the nature of which were ascertained as one methyl, one methoxyl, two methines, and seven quaternary carbon atoms. The ^{13}C NMR resonance at δ (ppm): 165.72, corresponding to C=O coupled with two methine carbon resonances at 133.62 and 123.63 having protons resonating at δ (ppm): 6.74 (d, $J = 16$ Hz, 1H) and 6.36 (d, $J = 16$ Hz, 1H,) confirmed the *trans*-double bond to be in conjugation with the carbonyl system. The six quaternary carbon signals δ (ppm): 83.05, 80.61, 71.58, 71.11, 64.71, and 58.20 indicated a triacetylenic group. A very high-field carbon resonance at δ (ppm): 4.7 (due to paramagnetic shift) forms one terminal end of the acetylenic chain. Presence of carbon resonance at 51.03 (with corresponding proton signal at 3.36 integrating for three protons) confirmed methoxyl group forming the second terminus of the chain. All this data on comparison with previous literature reports (Greger 1978) confirmed **4** to be *trans*-matricaria ester.

Compound **5** has been isolated as a yellow crystalline compound showing APCI mass at *m/z* 231.09 $[M + 1]^{+}$. IR absorption bands at 2217, 2140 (C≡C) and 1646 (C=C) cm^{-1} suggested the presence of ene–diyne unit. This was further supported by C NMR signals at δ (ppm): 79.61, 76.05, 69.81, and 64.68 corresponding to 4 acetylenic carbons. The ene–diyne unit was on one end attached to a very high-field methyl at δ (ppm): 4.62 and on the other end to a trisubstituted oxygenated olefin at δ (ppm): 164.92 and 85.83 [δ (ppm) 5.18 (br s, 1H)] which is because of strong inductive and conjugative effects of oxygen atom and diacetylene group. The quaternary carbon resonance at δ (ppm) 105.90 along with proton signals at δ 3.89 (m, 2H), 3.76 (d, $J = 2.5$ Hz, 1H), and 1.79-1.59 (6H, m) suggested the presence of a cyclic spiroacetal unit that is common in natural acetylenes (Liang et al. 2011). All this spectral data suggested compound **5** as diacetylenic spiroenol ether, namely (2E,3S,4R,5R)-3,4-epoxy-2-(2,4-hexadiynylidene)-1,6-dioxaspiro[4.5] decane which was confirmed on comparison with literature (Bohlmann et al. 1973).

Compound **6** has been isolated as a brown solid. Its mass, IR, and NMR spectral data was similar to that of **6**, but for the *J* value of the double bond protons (which was 11.2 Hz compared to 16 Hz in Compound **4**). This indication along with previous reports (Stavholt and Sorensen 1950) suggested that compound **6** was a *cis*-diastereomer of matricaria ester.

4.3 Biological Evaluation

4.3.1 *Bioactivity of the Extracts and Isolated Constituents Using SRB Assay*

The cytotoxic effect of hexane extracts of root and shoot as well as the isolates has been evaluated using Sulphorhodamine B (SRB) assay. All the human cancer cell lines (A-549, THP-1, PC-3, HCT-116) have been obtained from ATCC Sigma. Cells used were grown in RPMI-1640 medium containing 10 % Fetal bovine serum, 100 unit penicillin/100 μg Streptomycin per ml medium. Cells have been allowed to grow in carbon dioxide incubator (Thermoscientific USA) at 37°C with 98 % humidity and 5 % CO_2 gas environment in case of SRB assay. The SRB dye binds to the basic protein of cells that have been fixed to tissue culture plates by trichloroacetic acid (TCA). As the binding of SRB is stoichiometric, the amount of dye extracted from stained cells is directly proportional to the cell number. In the present case, all cell lines have been seeded in flat-bottomed 96-well plates were allowed to adhere overnight, and then media containing different samples (varying concentrations) were added. The plates have been assayed for 48 h. The cells have been fixed by adding 50 μl of ice-cold 50 % TCA to each well for 60 min. The plates have been washed five times in running tap water and stained with 100 μl per well SRB reagent (0.4 % w/v SRB) in 1 % acetic acid for thirty minutes. The plates have been washed five times in 1.0 % acetic acid to remove unbound SRB and allowed to dry overnight. SRB was solubilized with 100 μl per well 10 mM tris-buffer, shaken for 5 min, and the optical density has been measured at 570 nm (Kaur et al. 2005) (Fig. 4.1). However, the methanolic extracts have been found to be inactive. As a result isolation of the active principles responsible for such effect has been carried out and the isolates assayed for their cytotoxicity profile using SRB assay. From the results presented in Table 4.1, it is clear that except compound **1**, all others exhibited greater than 50 % inhibition at the tested concentration, so they were further screened to determine IC_{50} values. Almost the compounds displayed cytotoxity against all the four tested cancer cell lines. Ludartin (**2**), a highly active molecule displayed a broad spectrum cytotoxic profile against all the tested cancer cell lines, viz., human lung (A-549), leukemia (THP-1), prostate (PC-3), and colon (HCT-116) cells with corresponding IC_{50} values of 7.4, 3.1, 7.5 and 6.9 μM respectively. Compound **3** also showed potent cytotoxicity against lung cancer cell line (A-549) displaying IC_{50} of 5.9 μM.

4.3.2 *Bioactivity of Ludartin Using MTT Assay*

All the constituents have been further evaluated against Human dermal fibroblasts (CRL-1635) using MTT assay (Table 4.2). Owing to high cytotoxicity of Ludartin (**2**), it was further evaluated in MTT assay on Human epidermoid carcinoma

Table 4.1 Cytotoxicity profile of the isolated constituents using SRB assay

Compound	Conc (μM)	% Growth inhibition			
		Lung	Leukemia	Prostate	Colon
		A-549	THP-1	PC-3	HCT-116
1	50	68 ± 1.4	0	18 ± 0.4	0
	30	41 ± 1.7	0	16 ± 2.2	0
	20	30 ± 2.2	0	12 ± 1.3	0
	10	11 ± 0.3	0	5 ± 1.0	0
	5	10 ± 1.0	0	3 ± 1.7	0
	IC_{50}	37 ± 0.8	>50	>50	>50
2	50	97 ± 1.1	98 ± 1.3	97 ± 2.5	97 ± 1.1
	30	96 ± 1.0	90 ± 1.7	88 ± 2.0	97 ± 1.9
	20	77 ± 2.1	94 ± 2.0	70 ± 3.2	94 ± 2.1
	10	75 ± 0.4	92 ± 1.1	64 ± 1.9	91 ± 0.9
	5	34 ± 2.1	45 ± 1.7	39 ± 2.0	30 ± 0.4
	IC_{50}	7.4 ± 0.5	3.1 ± 0.3	7.5 ± 0.6	6.9 ± 0.5
3	50	95 ± 2.3	76 ± 2.7	20 ± 1.0	62 ± 1.6
	30	77 ± 0.5	55 ± 1.0	10 ± 1.7	43 ± 1.2
	20	75 ± 1.1	46 ± 2.1	07 ± 0.9	22 ± 3.1
	10	54 ± 2.0	34 ± 2.6	04 ± 0.8	11 ± 2.0
	5	52 ± 1.6	31 ± 1.8	00	09 ± 0.4
	IC_{50}	5.9 ± 0.4	21 ± 1.2	>50	41 ± 0.2
4	50	96 ± 1.2	97 ± 2.1	96 ± 0.5	99 ± 2.3
	30	64 ± 1.5	95 ± 2.2	89 ± 2.1	89 ± 1.1
	20	51 ± 1.3	53 ± 2.1	71 ± 1.8	58 ± 2.1
	10	17 ± 2.2	22 ± 1.9	47 ± 1.9	31 ± 2.8
	5	08 ± 2.7	10 ± 2.8	21 ± 2.2	16 ± 1.3
	IC_{50}	24 ± 1.1	21 ± 1.1	14 ± 2.8	18 ± 1.0
5	50	91 ± 0.8	93 ± 1.6	44 ± 2.0	86 ± 0.7
	30	73 ± 1.3	79 ± 1.7	31 ± 0.8	61 ± 2.3
	20	56 ± 1.5	60 ± 1.2	25 ± 1.1	55 ± 1.2
	10	23 ± 1.9	44 ± 1.1	17 ± 2.7	37 ± 2.0
	5	07 ± 1.0	10 ± 2.0	11 ± 2.1	22 ± 1.4
	IC_{50}	23 ± 0.8	19 ± 1.4	64 ± 3.1	18 ± 1.6
6	50	99 ± 1.2	98 ± 2.5	76 ± 1.5	86 ± 2.3
	30	78 ± 1.4	66 ± 2.2	59 ± 2.3	61 ± 3.0
	20	64 ± 1.1	58 ± 1.7	28 ± 2.0	55 ± 2.1
	10	47 ± 1.9	23 ± 1.9	15 ± 0.8	37 ± 0.5
	5	23 ± 2.3	10 ± 2.8	08 ± 1.1	22 ± 1.1
	IC_{50}	14 ± 1.4	22 ± 2.0	31 ± 0.6	18 ± 0.9

Table 4.2 Cytotoxicity of the isolates against human dermal fibroblasts (CRL-1635) by MTT assay

Compound	Conc. (μM)	% Cytotoxicity
1	10	9 ± 2.1
	25	Nil
	50	Nil
2	10	5 ± 0.2
	25	81 ± 1.6
	50	85 ± 1.8
3	10	6 ± 1.3
	25	2 ± 0.3
	50	2 ± 0.5
4	10	Nil
	25	Nil
	50	10 ± 0.3
5	10	7 ± 0.4
	25	2 ± 0.2
	50	Nil
6	10	9 ± 0.7
	25	18 ± 2.1
	50	13 ± 0.9

(A-431) and Mouse melanoma (B16F10) cell lines in a 96 well plate format. Ludartin (**2**) being the only member that displayed high degree of cytotoxicity was therefore further screened against human epidermoid carcinoma (A-431) and Mouse melanoma (B16F10) cell lines. Highest cytotoxicity was observed against mouse melanoma (B16F10) with IC_{50} of 6.6 μM followed by human epidermoid carcinoma (A-431) cell lines with IC_{50} of 19.0 μM (Table 4.3).

4.4 HPLC Analysis

Quantitative HPLC analysis has been performed on a Thermo-finnigan HPLC system equipped with a HPLC pump (P-4000), an auto sampler (AS-3000), a column oven, a Diode array detector (UV-6000 LP), vacuum membrane degasser (SCM 1000), and a system integrator (SN 4000) controlled by a ChromQuest 4.0 software which was used for data analysis and processing. Separation has been carried out on RP-C18 column (250 mm × 4.6 mm; particle size 5.0 mm; Merck, Germany) with column oven temperature of 25 °C. The flow rate has been adjusted at 0.5 ml/min and the injection volumes ranging from 2.0 to 12.0 μl. All the analysis was carried out at room temperature at a wavelength of 227 nm with run times ranging from 25 to 35 min.

A number of solvent systems have been tried to develop a separation method for the isolates from *Artemisia amygdalina* Decne extract solutions starting from pure methanol and constantly adding the aqueous phase. Since no resolution of

Table 4.3 Cytotoxicity profile of Ludartin (**2**) against Mouse melanoma (B16F10) and Human epidermoid carcinoma (A-431) cell lines

Compound	B16F10 (Mouse melanoma)		A-431 (Human epidermoid carcinoma)	
2	Conc (μM)	% cytotoxicity	Conc (μM)	% cytotoxicity
	1	Nil	1	12 ± 0.5
	2.5	19 ± 1.1	5.0	21 ± 1.5
	10	77 ± 2.6	10	59 ± 2.7
	20	76 ± 2.8	25	63 ± 1.5
	50	84 ± 3.2	50	72 ± 1.5
	75	88 ± 3.8	100	86 ± 1.9
	$IC_{50} = 6.6$ μM		$IC_{50} = 19.0$ μM	

compounds has been observed, the mobile phase has been changed to acetonitrile and constantly added the aqueous phase in increasing ratios up to acetonitrile:water. (70:30). Further addition of water (ACN:H_2O, 60:40 and 50:50) has resulted in very poor chromatographic resolution. Finally a simple isocratic-elution method comprising of acetonitrile:water (70:30) has been optimized to get the compounds resolved. Under the conditions of operation of the developed method optimum resolution, clear baseline separation with reasonable retention time and no tailing and fronting of peaks of the analytes has been observed. Maximally efficient detection has been observed at a fixed wavelength of 227 nm. Compounds **2**, **3**, **4**, **5**, and **6** were found to elute at 6.16, 5.20, 9.18, 7.75, and 6.79 min respectively with highly symmetric and well resolved peaks. The HPLC chromatograms are shown in Fig. 4.3, which depict a good separation of the peaks for all the analytes tested.

4.4.1 Linearity

The developed HPLC method has been developed to validate the linearity of the developed method. Solutions of Ludartin (**2**) (100–300 μg/ml), 5-hydroxy-6,7,3′,4′-tetramethoxyflavone (**3**) (20–120 μg/ml), *trans*-matricaria ester (**4**) (20–120 μg/ml), Diacetylenic spiroenol ether (**5**) (13–79 μg/ml), and *cis*-matricaria ester (**6**) (20–120 μg/ml) have been used.

The concentration range is generally choosen as per ICH guidelines (International Conference on Harmonization i.e., 70 and 130 % of the nominal concentration. Thus in a common way only one stock solution has been prepared and subsequently diluted to different concentration levels both above and below the nominal concentrations for each analyte. A 4.0 μl volume of standard solution has been injected and analyzed using the HPLC method as described above. Triplicate analysis for each analyte has been carried out. Calibration curves for all the compounds have been obtained using linear regression analysis. Good correlation coefficients ranging between 0.9962 and 0.9999 have been observed.

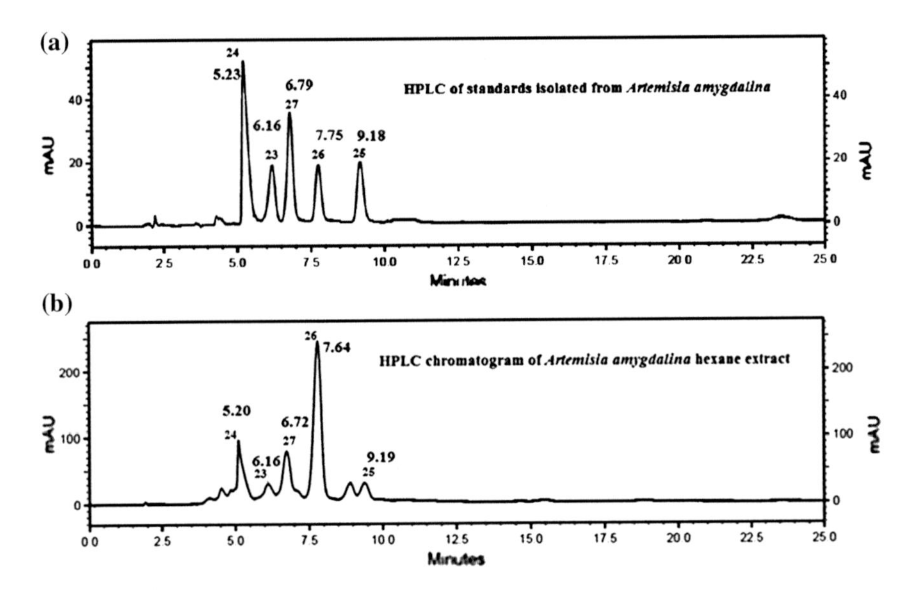

Fig. 4.3 **a** HPLC chromatogram of the standards isolated from *Artemisia amygdalina* Decne. **b** HPLC chromatogram of the extract solution of *Artemisia amygdalina* Decne

Other quantification parameters such as LOD and LOQ have also been calculated. LOD and LOQ of these cytotoxic constituents ranged between 0.0011 to 0.0506 and 0.0036 to 0.1533 μg/μl respectively, indicating that the developed method for *Artemisia amygdalina* Decne exhibited good sensitivity. All the results have been shown in Table 4.4

4.4.2 Accuracy

Accuracy has been evaluated by analyzing the low, mid and high concentrations of the analytes viz, ludartin (50, 100 and 150 μg/ml), 5-hydroxy-6,7,3′,4′-tetramethoxy flavone (30, 60 and 90 μg/ml), *Trans*-matricaria ester (30, 60 and 90 μg/ml), Diacetylenic spiroenol ether (26, 52 and 78 μg/ml), and *Cis*-matricaria ester (30, 60 and 90 μg/ml) and mixing with the extract solution of *Artemisia amygdalina* in 1:1 ratio (v/v). A 4.0 μl volume of standard solution has been injected and analyzed using the HPLC method as described above. The analysis was done in triplicates for all the samples. Sample recovery has been calculated as

$$\text{Recovery}(\%) = \frac{2 \times \text{Measured compound conc.} - \text{Compound conc. in } A.amygdalina \text{ extract}}{\text{Compounds theoretical concentration}} \times 100$$

Relative standard deviation (R.S.D) has been calculated as the standard deviation (S.D.) to the average value from the triplicate analysis. The results have been summed up in Table 4.5. The recovery rates of the analytes ranged between

Table 4.4 Parameters of quantification for the constituents from *A. amygdalina*

Analyte	Wavelength (nm)	Linear range (μg/ml)	Regression equation	r^2	LOD (μg/μl)	LOQ (μg/μl)
Ludartin (**2**)	227	100–300	Y = 2.32075e − 006X − 0.134127	0.9973	0.0506	0.1533
5-Hydroxy-6,7,3′,4′-tetramethoxy Flavone (**3**)	227	20–120	Y = 4.36e − 007X + 0.00264294	0.9999	0.0060	0.0181
Trans-matricaria ester (**4**)	227	20–120	Y = 2.72745e − 008X + 0.000234241	0.9998	0.0023	0.0072
Spiroenol ether (**5**)	227	13–79	Y = 2.60603e − 007X − 0.00409097	0.9962	0.0080	0.0242
Cis-Matricaria ester (**6**)	227	20–120	Y = 4.09939e − 008X + 0.000586977	0.9999	0.0011	0.0036

Table 4.5 Accuracy validation of HPLC method for the root and shoot constituents of *A. amygdalina*

Analyte	Spiked level (μg/ml)	Recovery (%)				
		1	2	3	Mean ± S.D.	R.S.D (%)
Ludartin (**2**)	50	103.15	105.42	103.16	103.90 ± 1.30	1.25
	100	96.42	98.33	101.16	98.63 ± 2.38	2.41
	150	95.48	103.22	100.61	99.77 ± 3.93	3.93
5-Hydroxy-6,7,3′,4′-tetramethoxy flavone (**3**)	30	103.22	105.64	106.60	105.15 ± 1.74	1.65
	60	98.78	97.33	99.00	98.37 ± 0.90	0.92
	90	99.43	100.51	102.13	100.69 ± 1.35	1.34
Trans-matricaria ester (**4**)	30	100.64	102.13	103.16	101.97 ± 1.26	1.23
	60	97.98	99.64	98.61	98.74 ± 0.83	0.84
	90	101.12	99.31	100.50	100.31 ± 0.91	0.90
Spiroenol ether (**5**)	26	97.86	98.32	98.99	98.39 ± 0.57	0.57
	52	99.14	100.15	98.16	99.15 ± 0.99	0.99
	78	100.12	100.56	102.13	100.93 ± 1.05	1.04
Cis-matricaria ester (**6**)	30	96.86	97.88	101.23	98.65 ± 2.28	2.31
	60	101.00	98.19	100.43	99.87 ± 1.48	1.48
	90	100.15	102.13	103.24	101.84 ± 1.56	1.53

98.37 ± 0.90 and 105.15 ± 1.74 % with relative standard deviations in the range of 0.57–3.93 % i.e., less than 5 % indicating high accuracy of the developed method.

4.4.3 Precision

To validate the precision of the developed method, theoretical concentrations of all the isolated compounds viz. Ludartin (100 μg/ml), 5-hydroxy-6,7,3´,4´-tetramethoxyflavone

Table 4.6 Precision validation of the HPLC method developed for quantification of *A. amygdalina*

Compound	Theort.Conc (μg/ml)	Measured concentrations (μg/ml)				Intraday Precision	Interday Precision
		Day 1	Day 2	Day 3		(%) CV	(%) CV
		M_1	M_2	M_3	TM		
2	100	98.97	100.72	101.83	100.50	1.93	0.90
3	60	60.49	60.79	63.00	61.42	3.43	1.02
4	60	61.49	60.03	62.73	61.42	3.23	1.15
5	52	52.35	50.40	51.54	51.43	3.15	0.54
6	60	59.12	61.29	58.37	59.92	1.15	3.08

(60 μg/ml), *Trans*-matricaria have been prepared. A 4.0 μl volume of standard solution has been injected and analyzed using the HPLC method as described above. Concentrations of all the compounds have been calculated using linear regression equation of the calibration curves of the isolated compounds. The experiment has been triplicate for each day and also per day over a 3-day period.

The limit of detection (LOD) and limit of quantification (LOQ) has been calculated using the equations

$$\mathrm{LOD} = 3.3\sigma/\mathrm{S} \quad \text{and} \quad \mathrm{LOQ} = 10\,\sigma/\mathrm{S}$$

where σ and S represent the standard deviation of response and the slope of the calibration curve, respectively.

Based on the triplicate analysis carried out each day and also per day over a 3 day period, the intra as well as interday precision levels for the developed method was analyzed. The results are presented in Table 4.6. Coefficient of variation for both the intraday and interday precision ranged between 1.15 to 3.43 % and 0.54 to 3.08 % respectively, calculated using single factor ANOVA. These results depicted a good precision of the developed analytical method.

4.4.4 Percentage of the Constituents

Using the developed method the concentration of constituents was determined on dry weight basis of plant material. The amount of isolated constituents present in *Artemisia amygdalina* Decne was found to be 0.3070 % (**2**), 0.0282 % (**3**), 0.012 % (**4**), 0.126 % (**5**), and 0.0711 % (**6**) on dry weight basis. Ludartin (**2**), the most active gauanolide isolated from the shoot part proved to be the most abundant constituent.

4.5 Conclusion

To date, this is the first study on the isolation, characterization, bioevaluation, HPLC quantification, and validation of chemical constituents of *Artemisia amygdalina* Decne. The findings have demonstrated the high cytotoxic potential of this plant in an in vitro manner, attributed mainly to the occurence of Ludartin (Position isomer of an approved antitumour agent, arglabin) screened via both SRB and MTT cytotoxicity assays. Further the presence of the highly bioactive compound, ludartin (**2**), in high concentrations (0.3 %), opens new avenues with regard to drug development. Because the developed method is simple, sensitive, selective, and repeatable, it can be extended for marker-based standardization of herbal formulations containing *Artemisia amygdalina* Decne and its use in pharmaceutical industries.

References

Bora KS, Sharma A (2011) Pharmaceutical Biol 49:101–109

Bhadane NR, Shafizadeh F (1975) Phytochemistry 1:2651–2653

Bohlmann F, Burkhardt T, Zdero C (1973) Naturally occurring acetylenes. Academic London

Efferth T (2007) Planta Med 73:299–309

Geissman TA, Griffin TS (1972) Phytochemistry 11:833–835

Greger H (1978) Phytochemistry 17:806

Harbone JB, Tomas-Barbera FA, Williams C, Gill M (1986) Phytochemistry 25:2811–2816

Kaur S, Kamboj S, Singh J, Saxena AK, Dhuna VA (2005) Biotechnol Lett 27:1815–1820

Kim JH, Kim HK, Jeon SB, Son KH, Kim EH, Kang SK, Sung ND, Kwon BM (2002) Tetrahedron Lett 43:6205–6208

Kwak JH, Jang WY, Zee OP, Lee RK (1997) Planta Med 63:474–476

Laid M, Hegazy MEF, Ahmed AA (2008) Phytochemistry Lett 85–88

Liang M, Fan G, Chun-Ping T, Chang-Qiang K, Xi-Qiang L, Andreas A, Yang Y (2011) Tetrahedron 2(67):3533–3539

Ruikar AD, Jadhav RB, Usha D, Rojatkar PSR, Puranik VG, Deshpande NR (2011) Helv Chim Acta 94:73–77

Smania JA, Delle FM, Smania EFA, Cuneo RS (1999) Int J Med Mushrooms 1:325–330

Sosa VE, Oberti JC, Gil RR, Ruveda EA, Goedken VL, Gutierrez AB, Herz W (1989) Phytochemistry 28:1925–1929

Stavholt K, Sorensen NA (1950) Acta Chem Scand 4:1567–1574

Tan RX, Lu H, Wolfender JL, Yu TT, Zheng WF, Yang LS, Hostettmann K (1998) Planta Med 65:64–67

Thomas OP, Ortet R, Prado S, Mourya E (2008) Phytochemistry 69:2961–2965

Wong H, Brown GD (2002) Phytochemistry 59:529–536

Chapter 5
Biological Profile of *Artemisia amygdalina*

Abstract This chapter deals with the biochemical pharmacology depicted by the plant *Artemisia amygdalina*. The important biological activities that have been reported till date include those of free radical scavenging potential of the in vitro raised and greenhouse acclimatized plants, anti-inflammatory, and immunomodulatory activity of the plant and the anti-diabetic and anti-hyperlipidemic effect of *A. amygdalina*. The methanolic extract of the in vitro grown and greenhouse acclimatized plants revealed the highest inhibitory activity, 92.11 and 91.2 % (IC_{50} = 26.06 μg mL^{-1}), respectively, against DPPH radical. Carrageenan paw edema model has been employed to study the potential of the plant extracts in inflammation in wistar rats. SRBC-specific haemagglutination-titer and DTH assays have been carried out in Balb/C mice for observing the effect of extracts on immune system. The methanolic fraction has been observed to have the maximum effect on the inhibition of paw edema formation with the inhibitory potential of 42.26 %, while in the immunomodulation studies the plant extracts have been found to have the immunosuppressant activity with methanolic fraction again showing the maximum potential for the suppression of both humoral (55.89 and 47.91 %) and cell-mediated immunity (62.27 and 57.21 %). Petroleum ether, ethyl acetate, methanol, and hydroethanolic extracts of *A. amygdalina* have been tested for their anti-diabetic potentials in diabetic rats. The hydroethanolic and methanolic extracts each at doses of 250 and 500 mg/kg b.w. have significantly reduced glucose levels in diabetic rats. The other biochemical parameters like cholesterol, triglycerides, low density lipoproteins (LDL), serum creatinine, serum glutamate pyruvate transaminase (SGPT), serum glutamate oxaloacetate transaminase (SGOT), and alkaline phosphatise (ALP), have been found to be reduced by the hydroethanolic and methanolic extracts. The extracts have also shown reduction in the feed and water consumption of diabetic rats when compared with the diabetic control.

Keywords Free radicals · DPPH assay · Deoxyribose assay · Anti-inflammatory · Immuno modulatory · Anti-diabetic

S.H. Lone et al., *Chemical and Pharmacological Perspective of* Artemisia amygdalina, SpringerBriefs in Pharmacology and Toxicology,
DOI 10.1007/978-3-319-25217-9_5

5.1 Free Radical Scavenging Activity Potential of *Artemisia amygdalina*

The free radical potential of the in vitro grown and greenhouse acclimatized plants has been reported by Rasool et al. (2013). Different assays like DPPH assay, Riboflavin Photo-oxidation assay, Riboflavin Photooxidation assay, Ferric thiocyanate (FTC) assay, Thiobarbituric acid (TBA) assay, Post-mitochondrial supernatant (PMS) assay, and DNA–damage assay have been carried out. In DPPH assay, the methanolic extract of the in vitro grown and greenhouse acclimatized plants revealed the highest inhibitory activity of about 92.11 and 91.2 % (IC_{50} 26.06 μg mL^{-1}), respectively, against DPPH radical. Vitamin C showed 77.93 % (IC_{50} = 16.37 μg mL^{-1}) inhibitory activity (Fig. 5.1 and Table 5.1).

In case of the Riboflavin Photooxidation assay, ethyl acetate extract of in vitro grown and the aqueous extract of greenhouse acclimatized plants showed 89.53 % (IC_{50} = 16.2 μg mL^{-1}) and 96.32 % (IC_{50} = 22.04 μg mL^{-1}), respectively, against superoxide radicals. Vitamin C showed 92.82 % (IC_{50} = 4.2 μg mL^{-1}) inhibitory activity (Fig. 5.2 and Table 5.1). However, in methanolic extract of in

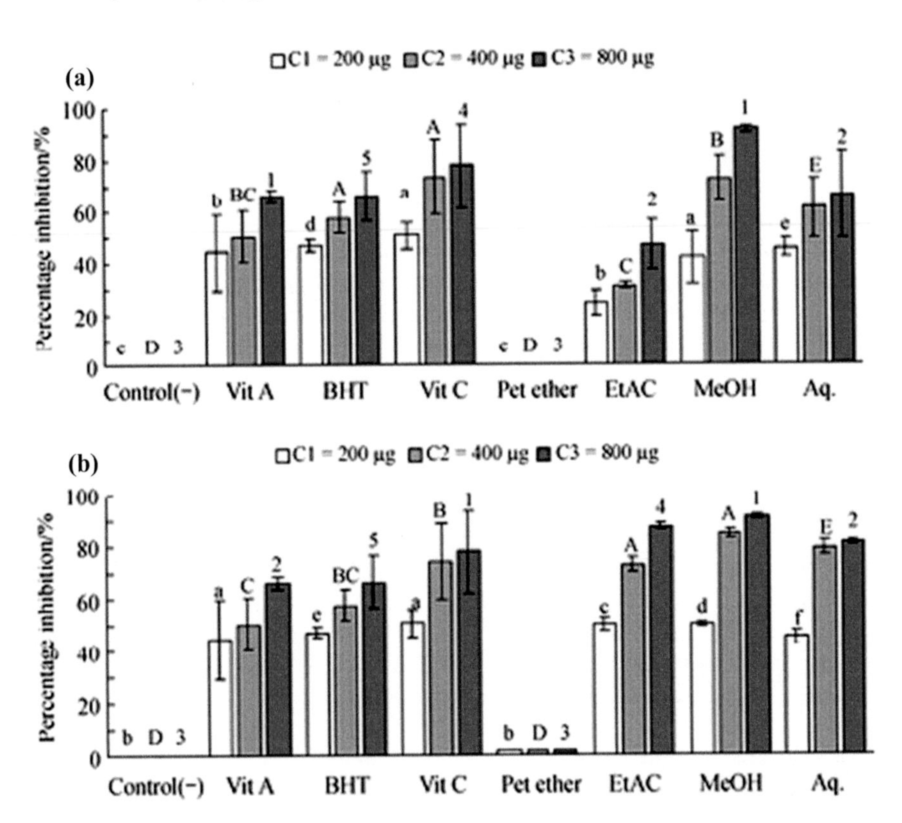

Fig. 5.1 DPPH radical activity of the **a** in vitro grown **b** greenhouse acclimatised plants of *Artemisia amygdalina*

Table 5.1 IC_{50} values of petroleum ether (P), ethyl acetate (E), methanol (M), and aqueous (W) extracts of in vitro grown (T) and greenhouse raised plants (G) of *A. amygdalina* ($n = 3$)

Sample	IC_{50} ($\mu g\ mL^{-1}$)					
	DPPH assay	Riboflavin photooxidation assay	Deoxyribose assay	FTC assay	TBA assay	PMS assay
VIT A	185.20	8.13	64.86	26.23	52.41	127.86
BHT	120	4.67	117.83	28.95	53.61	159.80
VIT C	16.37	4.2	212.86	28.31	110	180.67
PT	–	16.96	243.2	27.97	25.34	212.73
PG	–	15	33.36	25.46	28.62	959.16
ET	444	16.2	178.27	29.45	54.92	383.6
EG	62.47	35.96	25.06	20.58	14.6	–
MT	114.4	25.22	73.86	29.84	42.50	108.14
MG	26.06	18.69	13.96	31.57	41.43	82.11
WT	112.18	27.94	129.82	30.81	24.98	98
WG	58.11	22.04	32.17	33	26.93	124.02

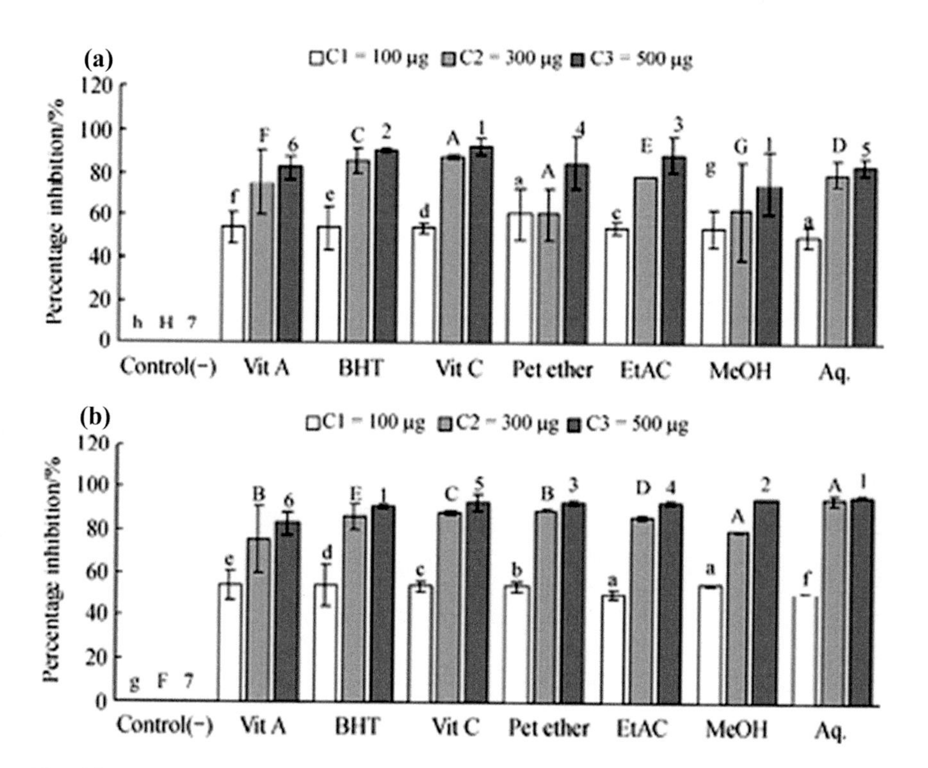

Fig. 5.2 Riboflavin photooxidation assay of **a** in vitro raised and **b** greenhouse acclimatized plants of *Artemisia amygdalina* Decne

vitro grown and acclimatized plants revealed higher percentage inhibitions, 46.17 and 69.33 % (IC_{50} = 13.96 µg mL^{-1}), respectively, against hydroxyl radicals generated by Fentons reaction. Vitamin E showed 50.13 % (IC_{50} = 64.86 µg mL^{-1}) inhibitory activity (Fig. 5.2 and Table 5.1).

In case of FTC assay, the ethyl acetate extract of in vitro grown and the petroleum ether extract of the acclimatized plants revealed 76.57 % (IC_{50} = 29.45 µg mL^{-1}) and 88.28 % inhibitions (IC_{50} = 25.46 µg mL^{-1}), respectively, against lipid peroxyl radicals generated by oxidation of linoleic acid. Vitamin E showed 79.69 % (IC_{50} = 26.23 µg mL^{-1}) of inhibitory activity (Fig. 5.3 and Table 5.1).

In TBA assay, petroleum ether extract of the in vitro raised and the aqueous extract of the greenhouse acclimatized plants revealed inhibitory activity, 64.1 % (IC_{50} = 25.34 µg mL^{-1}) and 64.15 % (IC_{50} = 26.93 µg mL^{-1}), respectively, against TBA reactive species. Vitamin E showed 42.3 % (IC_{50} = 52.41 µg mL^{-1}) inhibitory activity (Fig. 5.4 and Table 5.1).

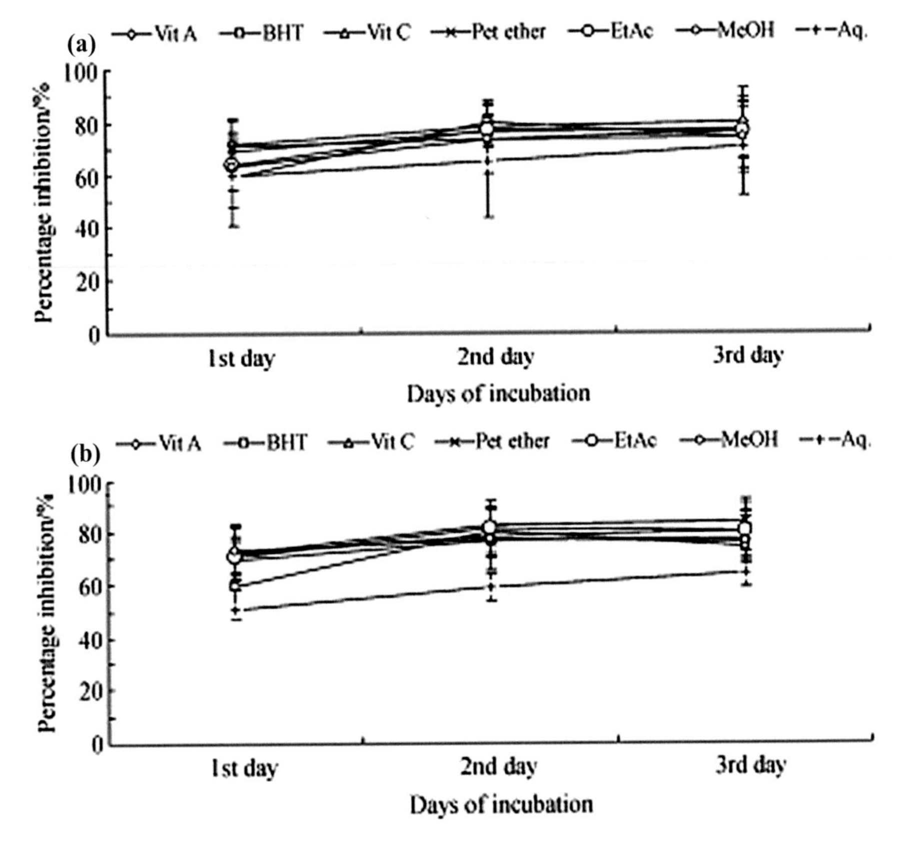

Fig. 5.3 Ferric thiocyanate assay of **a** in vitro raised and **b** greenhouse acclimatized plants of *Artemisia amygdalina* Decne

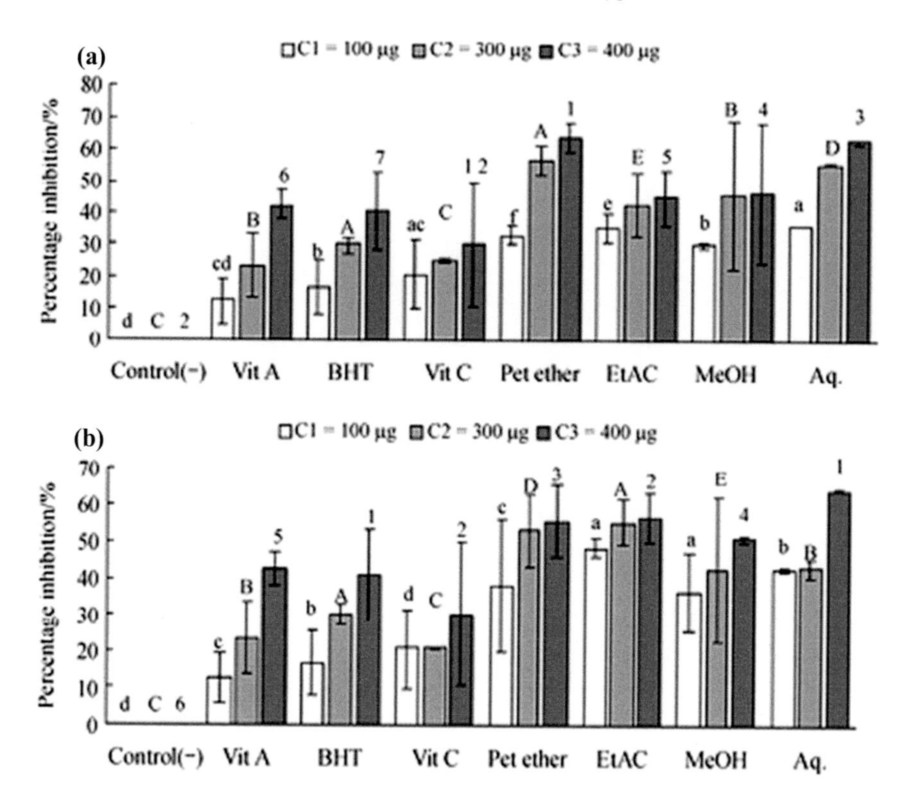

Fig. 5.4 Thiobarbituric acid assay of **a** in vitro raised and **b** greenhouse acclimatized plants of *Artemisia amygdalina* Decne

The PMS assay, when carried out revealed that the methanolic extract of in vitro raised and greenhouse acclimatized plants have more activity than other solvent extracts, 52.86 % (IC_{50} = 108.14 µg mL^{-1}) and 67.04 % (IC_{50} = 82.11 µg mL^{-1}), respectively. Vitamin E showed 48.96 % (IC_{50} = 127.86 µg mL^{-1}) inhibitory activity (Fig. 5.5 and Table 5.1). Further, DNA damage was estimated using agarose gel assay. It has been found that damage to DNA has been reduced with an increase in concentrations of plant extracts (1000 µg mL^{-1}). Higher concentration of in vitro raised and greenhouse acclimatized plants methanolic extracts displayed DNA protective activity (Fig. 5.6).

5.2 Anti-inflammatory and Immunomodulatory Potential of *Artemisia amygdalina*

The anti-inflammatory and then the immunomodulatory potential of this plant have been reported by Mubashir et al. (2013). Carrageenan-induced paw edema has been carried out to study the anti-inflammatory potential of *A. amygdalina*

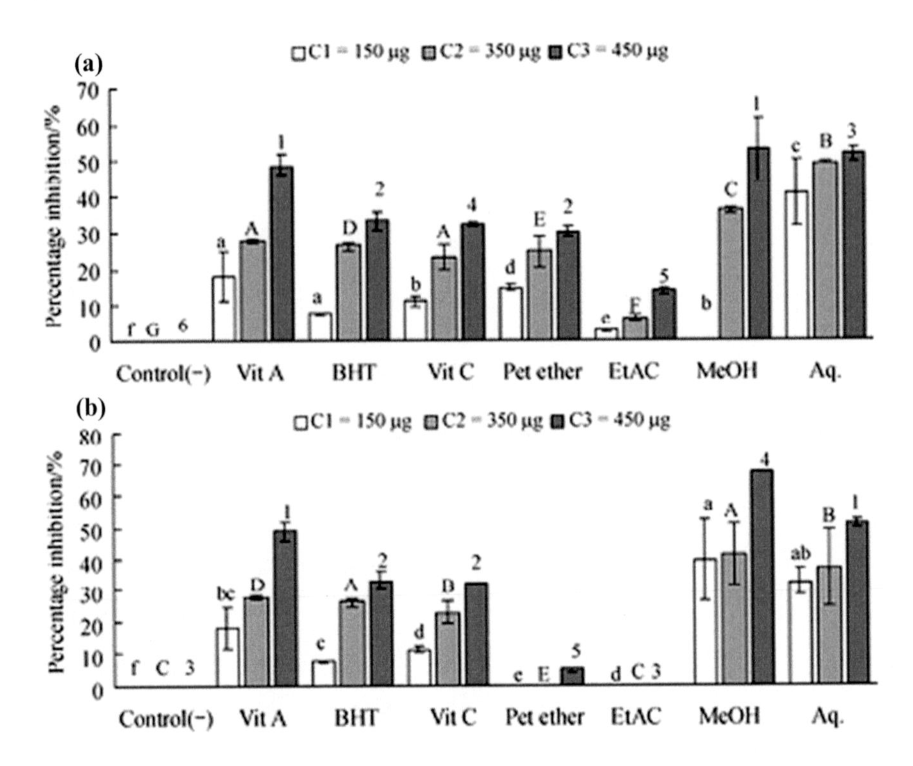

Fig. 5.5 Post supernatant mitochondrial assay of **a** in vitro raised and **b** greenhouse acclimatized plants of *Artemisia amygdalina* Decne

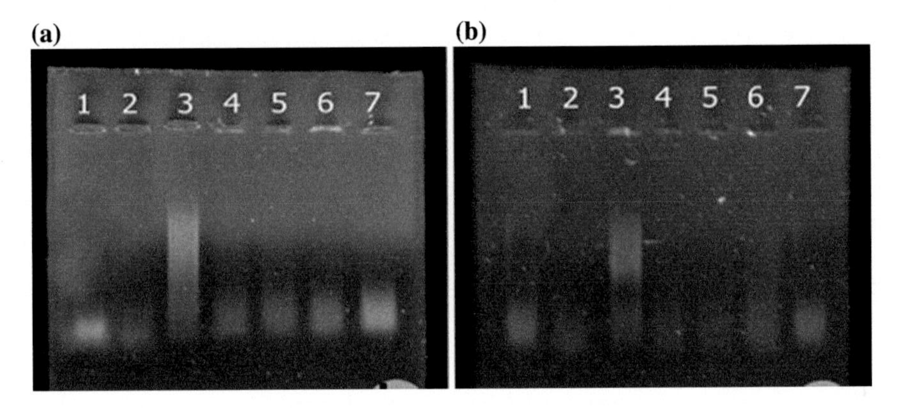

Fig. 5.6 Agarose gel electrophoresis for DNA damage inhibition assay by *A. amygdalina* methanolic extracts. **a** Tissue culture raised plant; **b** Greenhouse grown plant

Decne. To study the immunomodulatory potential of *A. amygdalina* Decne, SRBC-specific haemagglutination titer and DTH assays have been carried out in Balb/C mice.

5.2.1 Carrageenan-Induced Paw Edema

Carrageenan-induced paw edema model (Winter et al. 1962) has been utilized to assess the acute anti-inflammatory potential of the *A. amygdalina* extract solutions. Inflammation is a process that essentially preserves and maintains the integrity of the organism in the event of physical, chemical, and infectious damages. Very often, the inflammatory response to severe lesions erroneously damages normal tissues (Lunardelli et al. 2006). The injection of carrageenan into the hind paw of rats elicits an acute inflammatory response characterized by accumulation of fluid (edema). During acute inflammation, serum proteins and leukocytes migrate to areas of tissue injury. Recruitment of cells to inflammatory sites is dependent on the release of vasoactive and chemotactic factors that increase regional blood flow and microvascular permeability and promote the migration of leukocytes from the intravascular space into the tissues (Suffredini et al. 1999). In the current experiment by Mubashir et al. (2013), animals have been divided into six groups ($n = 4$). Group I served as control, rats in groups II–V were administered with plant extracts, and group VI was used as positive control. All test samples were given orally, 45 min before carrageenan injection The different extracts tested orally for anti-inflammatory activity at a dose of 250 mg/kg showed decrease in the paw edema after 4 h. Methanolic extract of *A. amygdalina* has shown maximum inhibition in paw edema (45.26 %) as compared to control followed by aqueous (29.47 %), ethyl acetate (18.95 %), and petroleum ether (3.16 %) extracts (Table 5.2). The inhibition in the paw edema of standard group observed was 52.63 %. The % inhibition by methanolic and standard drug is much comparable and is statistically significant at $P < 0.05$, while the results obtained in the petroleum ether-treated group are almost same as that of the control group. Diclofenac, however, served as a positive control in this assay.

Table 5.2 Effect of different extracts of *Artemisia amygdalina* D on Carrageenin-induced paw edema in rats (mean ± SE) ($n = 4$)

S. No	Groups	Dose mg/Kg	Initial paw vol ml	Paw vol after 4 h ml	Edema (4 h)	% Inhibition (4 h)
1	Control	NS	0.95 ± 0.03	1.90 ± 0.04	0.95 ± 0.03	–
2	AMT	250	1.00 ± 0.00	1.52 ± 0.05	0.52 ± 0.05	45.26
3	AET	250	0.97 ± 0.05	1.75 ± 0.06	0.77 ± 0.05	18.95
4	APT	250	0.95 ± 0.04	1.87 ± 0.09	0.92 ± 0.07	3.16
5	AAQ	250	0.92 ± 0.02	1.60 ± 0.04	0.67 ± 0.05	29.47
6	Diclo.	20	1.07 ± 0.07	1.52 ± 0.05	0.45 ± 0.03	52.63

5.2.2 Humoral Antibody Titer

Humoral antibody assay has been carried out by Mubashir et al. (2013) in Balb mice. In a usual procedure, the mice have been divided into six groups, each consisting of five animals. Mice in group I (control) were given 1 % tween-20, 0.2 ml/mouse for 14 days. Mice in group II–V have been given test samples (plant extracts of *A. amygdalina*), 100 mg/kg b.w. (orally) for 14 days. Mice in group VI have been given cyclophosphamide 50 mg/kg on day 1 and continued for 14 days. The animals have been immunized by injecting 200 μL of 5 × 109 SRBCs/mL, intraperitoneally (i.p.) on day 1. Blood samples have been collected in microliter-tubes from individual animals of all the groups by retro-orbital vein puncture on day 7 and day 14. The blood samples have been centrifuged, and the serum separated. Then, haemagglutination primary and secondary titers have been performed as per the protocols (Bagwat et al. 2010; Thakur et al. 2011). Various fractions of *A. amygdalina* have been shown to decrease the effect on primary and secondary antibody formation compared to control. But, methanolic fraction has produced maximum decrease in humoral response followed by ethyl acetate and aqueous fractions at a dose of 100 mg/kg (Table 5.3). The observed decrease in primary and secondary antibody titers in methanolic fraction has been found to be 55.89 and 47.91 %, while that of ethyl acetate fraction as 48.53 and 37.5 %, and aqueous fraction as 41.17 and 38.89 %, respectively. Petroleum ether fraction has shown decrease in primary response (11.76 %), but a slight increase in secondary response (2.77 %). The immunosuppressant activity shown by the three fractions is much comparable to cyclophosphamide 50 mg/kg used as a standard drug inducing 38.23 and 52.77 %, decrease in primary and secondary titers, thus indicating that these fractions of *A. amygdalina* significantly ($P < 0.05$) inhibit antibody formation (Fig. 5.7).

Table 5.3 Effect of different extracts of *A. amygdalina* D on haemagglutination titer in mice (mean ± SE) ($n = 5$)

Humoral response			
Groups	Dose (mg/Kg)	Primary titer	Secondary titer
Control	SRBC	6.8 ± 0.38	7.2 ± 0.49
Methanolic fraction	100	3.0 ± 0.41	3.75 ± 0.43
Ethyl acetate fraction	100	3.5 ± 0.29	4.5 ± 0.26
Petroleum ether fraction	100	6.0 ± 0.45	7.4 ± 0.24
Aqueous fraction	100	4.0 ± 0.89	4.4 ± 0.24
Cyclophosphamide	50	4.2 ± 0.58	3.4 ± 0.24

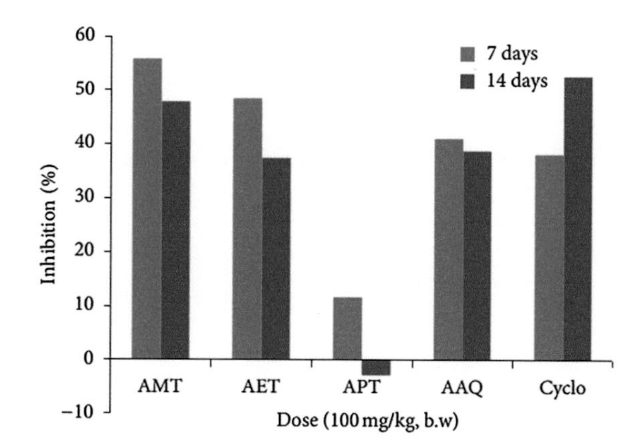

Fig. 5.7 Graph showing percentage inhibition of primary and secondary responses in Balb/C mice by different fractions of *Artemisia amygdalina* D

5.2.3 Delayed-Type Hypersensitivity

This assay has been carried out to measure the immunosuppressant potential of *A. amygdalina* extracts. Like humoral assay animals have been divided into six groups of five each. Group I served as sensitized control, as in humoral antibody titer. Mice in group II–V have been administered extracts of *A. amygdalina* after SRBCs' sensitization and once daily for seven days. Cyclophosphamide (50 mg/kg) has been administered as standard T-cell suppressor (group VI). The mice were challenged by injecting the same amount of SRBCs intradermally into the right hind footpad, whereas left hind footpad served as control (Mangathayaru et al. 2009; Jayathirtha and Mishra 2004). The T-cell-mediated DTH response to sheep RBC has shown a decrease in the paw volume in test groups as compared to control group. Out of the different test groups, methanolic fraction-treated group has shown maximum decrease in paw volume which is much comparable to the results seen in cyclophosphamide treated group, used as standard drug (Table 5.4). The percentage decrease in edema formation in methanolic fraction-treated group and standard group observed was found to be 62.27 and 68.65 % and 57.21 and 76.71 % after 24 and 48 h, respectively, Fig. 5.8. The results observed in test groups were much significant ($P < 0.05$) compared to the control group.

Table 5.4 Effect of different extracts of *Artemisia amygdalina* on delayed-type hypersensitivity response in mice (mean ± S.D)

DTH assay

Groups	Dose (mg/Kg)	24 h paw thickness	48 h Paw thickness
Control	SRBC	0.6825 ± 0.04	0.335 ± 0.021
Methanolic fraction	100	0.2575 ± 0.02	0.105 ± 0.012
Ethyl acetate fraction	100	0.386 ± 0.04	0.21 ± 0.023
Petroleum ether fraction	100	0.52 ± 0.04	0.306 ± 0.021
Aqueous fraction	100	0.368 ± 0.032	0.222 ± 0.027
Cyclophosphamide	50	0.292 ± 0.032	0.078 ± 0.006

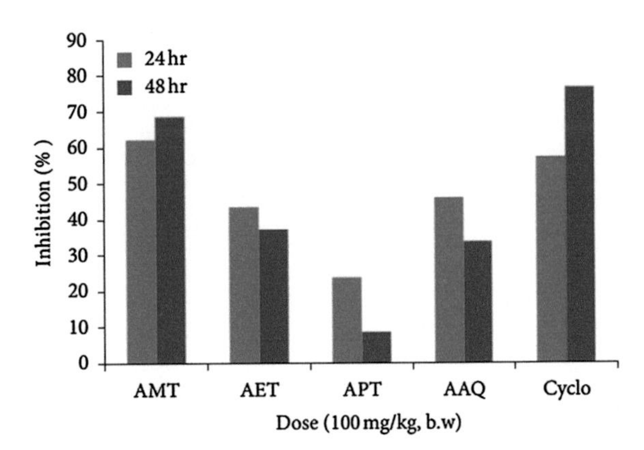

Fig. 5.8 Graph showing 24 and 48 h % age inhibition of cell-mediated response in mice by different fractions of *Artemisia amygdalina* D

5.2.4 Anti-diabetic Potential of Artemisia amygdalina

Based on the folklore claims of the genus *Artemisia* to be used for the treatment of diabetes, Gazafar et al. (2014) exploited the anti-diabetic and antihyperlipidemic effects of *A. amygdalina*. Various extracts prepared from the mentioned plant have been used for the study. Petroleum ether, ethyl acetate, methanol, and hydroethanolic extracts of *A. amygdalina* are some of the extracts tested for their anti-diabetic potentials in diabetic rats. The effect of extracts has been observed by checking the biochemical, physiological, and histopathological parameters in diabetic rats. The hydroethanolic and methanolic extracts, each at doses of 250 and 500 mg/kg b.w. significantly reduced glucose levels in diabetic rats. The other biochemical parameters like cholesterol, triglycerides, low density lipoproteins (LDL), serum creatinine, serum glutamate pyruvate transaminase (SGPT), serum glutamate oxaloacetate transaminase (SGOT), and alkaline phosphatase (ALP), were found to be reduced by the hydroethanolic and methanolic extracts. The extracts also showed reduction in the feed and water consumption of diabetic rats when compared with the diabetic control. The histopathological results of treated groups showed the regenerative/protective effect on β-cells of pancreas in diabetic rats. The current study revealed the anti-diabetic potential of *A. amygdalina* being effective in hyperglycemia and that it can effectively protect against other metabolic aberrations caused by diabetes in rats, which seems to validate its therapeutic traditional use.

In a typical experimental methodology, various extracts of *A. amygdalina* have been prepared, viz., petroleum ether, ethyl acetate, methanol, and hydroalcoholic extracts. Experimental animals that have been used in the experiment include Albino Wistar rats weighing 120–150 g of either sex. They have been fed a standard rat pellet and water from reverse osmosis purifier (Kent). Research on animals has been conducted in accordance with the guidelines of the Committee for the Purpose of Control and Supervision of Experiments on Animals (CPCSEA) as the institute has CPCSEA registration (Reg. number 927/GO/c/06/CPCSEA). The experimental protocol has been approved by the Institutional Animal Ethics Committee of the Regional Research Institute of Unani Medicine, Srinagar, Jammu and Kashmir, India.

5.2.5 Acute Oral Toxicity

Acute oral toxicity study has been performed for the extracts of *A. amygdalina* according to the guidelines of the organisation for Economic Cooperation and Development (OECD 2008). The rats have been kept on fasting overnight, being provided only water prior to oral dosing. Then the extract has been administered orally at different dose levels, i.e., 100, 200, 500, 1000, 1500, and 2000 mg/kg of body weight. The rats have been observed continuously for 24 h for behavioural and any adverse change and thereafter for any lethality. The extracts were found to be safe up to the dose level of 2000 mg/kg of body weight in rats. The extracts did not induce any toxicological effect in any rat. There was no lethality found by oral administration of any extracts of *A. amygdalina.*

5.2.6 Antihyperglycaemic Effect of *Artemisia amygdalina*

The effect of extracts of *A. amygdalina* and glibenclamide on serum glucose levels in normal, diabetic, and extract treated rats is presented in Table 5.5. The highest percent variation in fasting glucose levels were shown by hydroethanolic extract (54.07 %), followed by methanolic extract (43.93 %). The standard reference drug glibenclamide (3 mg/kg b.w.) was found to decrease fasting glucose levels by 48.09 % after 14 days of treatment. The other two extracts, i.e., petroleum ether and ethyl acetate, also showed inhibitory effect on glucose levels but were not considered for further studies. The highly bioactive fractions were tested for dose dependence and were observed to show increased antihyperglycaemic

Table 5.5 Effect of extract's of *Artemisia amygdalina* and glibenclamide on fasting blood glucose levels of rats

Groups	Serum glucose (mg/dL)			
	0th day	7th day	14th day	% variation
Normal	82.32 ± 0.99	85.43 ± 1.39	83.57 ± 2.22	−1.51
Diabetic control	360.14 ± 5.06	387.58 ± 3.28	410.13 ± 2.48	−13.88
Pet ether extract (500 mg/kg b.w.)	350.55 ± 3.26	309.87 ± 2.85	266.57 ± 4.10	23.95
Hydroethanolic extract (500 mg/kg b.w.)	386.47 ± 5.51	232.51 ± 3.82**	177.49 ± 2.00**	54.07
Methanolic extract (500 mg/kg b.w.)	359.54 ± 3.33	284.65 ± 2.83**	201.59 ± 3.38**	43.93
Ethyl acetate extract (500 mg/kg b.w.)	371.35 ± 2.92	325.43 ± 6.43	260.53 ± 2.94	29.84
Glibenclamide (3 mg/kg b.w.)	365.9 ± 3.97	248.78 ± 4.16**	189.92 ± 3.07**	48.09

The values are expressed as mean ± SEM; $n = 5$ in each group
**$P < 0.01$ as compared with normal at the same time (one-way ANOVA followed by Dunnett's multiple comparison test)

Table 5.6 Dose-dependent effect of hydroethanolic and methanolic extracts of *Artemisia amygdalina* and glibenclamide on fasting blood glucose levels of rats

Groups	Serum glucose (mg/dL)			
	0th day	7th day	14th day	% variation
Normal	85.66 ± 1.09	84.86 ± 1.13	84.12 ± 2.25	1.80
Diabetic control	375.41 ± 4.62	390.82 ± 4.75	419.52 ± 3.94	−11.75
Hydroethanolic extract (50 mg/kg b.w.)	380.76 ± 5.24	366.27 ± 4.55	352.78 ± 5.84	7.35
Hydroethanolic extract (100 mg/kg b.w.)	355.15 ± 4.48	330.85 ± 6.3	288.54 ± 3.96	18.76
Hydroethanolic extract (250 mg/kg b.w.)	370.59 ± 5.23	290.36 ± 6.34**	238.45 ± 7.46**	35.66
Hydroethanolic extract (500 mg/kg b.w.)	377.81 ± 5.45	227.55 ± 4.36**	162.92 ± 3.85**	56.88
Methanolic extract (50 mg/kg b.w.)	358.96 ± 6.94	350.12 ± 7.24	344.4 ± 5.74	4.06
Methanolic extract (100 mg/kg b.w.)	366.85 ± 5.89	340.53 ± 7.84	328.12 ± 4.85	10.56
Methanolic extract (250 mg/kg b.w.)	390.61 ± 6.37	358.24 ± 5.67	295.48 ± 6.13	24.35
Methanolic extract (500 mg/kg b.w.)	365.37 ± 8.7	278.42 ± 4.75**	200.64 ± 5.82**	45.09
Glibenclamide (3 mg/kg b.w.)	372.27 ± 4.86	258.87 ± 5.64**	195.25 ± 4.37**	47.55

The values are expressed as mean ± SEM; $n = 5$ in each group
**$P < 0.01$ as compared with normal at the same time (one-way ANOVA followed by Dunnett's multiple comparison test)

activity with increase in dose. The hydroethanolic fraction showed a more significant effect in decreasing blood glucose levels than the methanolic fraction and the maximum % variation observed at 500 mg/kg b.w. was 56.88 while the % variation at the same dose in methanolic fraction was 45.09 (Table 5.6).

5.2.7 *Effect of Extracts of Artemisia amygdalina and Glibenclamide on Various Biochemical Parameters in Rats*

The extracts of *A. amygdalina* hydroethanolic extract (500 mg/kg b.w.) and methanolic extract (500 mg/kg b.w.) significantly lowered the levels of cholesterol, triglycerides, LDL, and creatinine in diabetic rats when compared with the diabetic control group (Table 5.7). Total protein was found to be lowered in diabetic control group, while it was found to be elevated in the extract- and glibenclamide-treated diabetic rats. The extracts were also shown to significantly lower the enzymatic activity of liver marker enzymes (SGPT, SGOT, and ALP) in diabetic rats as represented in Table 5.8.

Table 5.7 Effect of extracts of *Artemisia amygdalina* and glibenclamide on various biochemical parameters in rats

Groups	Cholestrol (mg/dL)	Triglycerides (mg/dL)	LDL (mg/dL)	Protien (g/dL)	Creatinine (mg/dL)
Normal	50.59 ± 2.77	62.83 ± 2.46	83.35 ± 3.22	7.11 ± 0.30	0.69 ± 0.05
Diabetic control	149.33 ± 6.93	120.83 ± 4.96	189.38 ± 5.61	4.20 ± 0.44	1.98 ± 0.09
Hydroethanolic extract (250 mg/kg b.w.)	84.32 ± 3.67	95.74 ± 5.54	132.45 ± 3.35	5.75 ± 0.29	1.18 ± 0.05
Hydroethanolic extract (500 mg/kg b.w.)	64.83 ± 2.19**	75.66 ± 4.87**	101.36 ± 2.95**	6.12 ± 0.32	0.88 ± 0.08**
Methanolic extract (250 mg/kg b.w.)	88.12 ± 3.79	98.27 ± 5.28	142.77 ± 4.23	5.15 ± 0.26	1.07 ± 0.06
Methanolic extract (250 mg/kg b.w.)	73.35 ± 2.92**	84.43 ± 6.44**	128.54 ± 3.46**	6.67 ± 0.28**	0.97 ± 0.07**
Glibenclamide (3 mg/kg b.w.)	60.83 ± 2.73**	78.00 ± 2.73**	92.37 ± 3.67**	7.03 ± 0.21**	0.78 ± 0.04**

The values are expressed as mean ± sem; $n = 5$ in each group
**$P < 0.01$ as compared with diabetic control at the same time (one-way anova followed by dunnett's multiple comparison test)

Table 5.8 Effect of extracts of *Artemisia amygdalina* and glibenclamide on liver marker enzymes of streptozotocin induced diabetic rats

Groups	SGOT (U/L)	SGPT (U/L)	ALP (U/L)
Normal	115.00 ± 5.14	95.23 ± 6.75	175.67 ± 8.68
Diabetic control	243.00 ± 9.57	209.16 ± 6.35	296.17 ± 7.85
Hydroethanolic extract (250 mg/kg b.w.)	164.31 ± 6.53	148.76 ± 7.69	232.17 ± 6.82
Hydroethanolic extract (500 mg/kg b.w.)	129.66 ± 8.08**	112.83 ± 7.11**	201.66 ± 7.94**
Methanolic extract (250 mg/kg b.w.)	174.76 ± 4.83	177.62 ± 7.71	229.77 ± 6.18
Methanolic extract (500 mg/kg b.w.)	138.53 ± 2.94**	121.85 ± 6.69**	219.53 ± 7.37
Glibenclamide (3 mg/kg b.w.)	119.50 ± 4.43**	103.73 ± 3.00**	198.50 ± 7.14**

The values are expressed as mean ± SEM; $n = 5$ in each group
$^{**}P < 0.01$ as compared with diabetic control at the same time (one-way ANOVA followed by Dunnett's multiple comparison test)

5.2.8 Effect of Extracts of Artemisia amygdalina and Glibenclamide on Feed Consumption and Water Consumption in Rats

The extract treated rats (hydroethanolic and methanolic extract at a dose level of 500 mg/kg b.w.) and glibenclamide treated rats (3 mg/kg b.w.) significantly overcame the symptoms of diabetes, i.e., polyphagia and polydipsia. The extract- and glibenclamide-treated rats consumed less water and feed when compared with the diabetic control ones. Table 5.9 shows the effect of extracts and glibenclamide on feed and water consumption in rats.

5.2.9 Effect of Extracts of Artemisia amygdalina and Glibenclamide on Body Weight in Rats

The body weight of rats belonging to diabetic control group was drastically decreased upon the induction of diabetes. The extract- and glibenclamide- treated rats were found to gain body weight significantly when compared with the diabetic control group as shown in Table 5.10.

5.2.10 Histopathology

Administration of streptozotocin decreased the number of β-cells. With the result, the observed mean pancreatic weight of untreated diabetic group was less compared to the mean weight of pancreas of normal (non-diabetic) and diabetic treated

Table 5.9 Effect of the extracts of *Artemisia amygdalina* and glibenclamide on feed intake and fluid intake of the rats

Groups	Water consumption (mL/day)	Food consumption (g/day)
Normal (water)	24 ± 4.02	19.34 ± 2.36
Diabetic (water)	65 ± 7.60	29.67 ± 1.23
Hydroethanolic extract (250 mg/Kg b.w.)	46 ± 5.37	25.33 ± 3.30
Hydroethanolic extract (500 mg/kg b.w.)	39 ± 7.16**	23.56 ± 3.42**
Methanol extract (250 mg/kg b.w.)	49 ± 4.02	24.24 ± 2.60
Methanol extract (500 mg/kg b.w.)	44 ± 4.92**	22.87 ± 2.01**
Glibenclamide (3 mg/kg b.w.)	44 ± 6.71**	22.33 ± 1.23**

The values are expressed as mean ± SEM; $n = 5$ in each group
**$P < 0.01$ as compared with diabetic control at the same time (one-way ANOVA followed by Dunnett's multiple comparison test)

Table 5.10 Effect of extracts of *Artemisia amygdalina* and glibenclamide on body weight in streptozotocin induced-diabetic rats

Groups	Body weight (g)			
	0th day	7th day	14th day	% variation
Normal	120 ± 1.79	135 ± 3.13	155 ± 2.68	29.16
Diabetic	125 ± 1.34	118 ± 1.79	107 ± 3.13	−14.4
Hydroethanolic extract (250 mg/kg b.w.)	127 ± 2.24	132 ± 2.24	139 ± 3.58**	9.44
Hydroethanolic extract (500 mg/Kg b.w.)	122 ± 2.68	128 ± 3.13	135 ± 2.24**	10.65
Methanolic extract (250 mg/Kg b.w.)	125 ± 3.13	130 ± 3.58	137 ± 4.92**	9.6
Methanolic extract (500 mg/Kg b.w.)	123 ± 2.24	130 ± 3.58	140 ± 5.37**	13.82
Glibenclamide (3 mg/kg b.w.)	125 ± 1.79	133 ± 2.68	142 ± 3.13**	13.6

The values are expressed as mean ± SEM; $n = 5$ in each group
**$P < 0.01$ as compared with diabetic control at the same time (one-way ANOVA followed by Dunnett's multiple comparison test)

groups. The sections from the untreated diabetic group demonstrated shrunken islets of Langerhans with degenerative necrosis. There was an increased vacuolation (Fig. 5.10). In the extract treated rats (Figs. 5.9, 5.10, 5.11, 5.12, 5.13, 5.14 and 5.15) islets of Langerhans appeared less shrunken as compared to those of untreated diabetic group. The histological appearance of the pancreatic islet cells of the control was normal (Fig. 5.10). The morphology of the pancreas of *A. amygdalina* extracts (hydroethanolic and methanolic) treated diabetic rats revealed remarkable improvement in the islets of Langerhans. There was an increase in the islet cellular density, with an increase in granulation, and vacuolation was reduced or absent in many islets (Figs. 5.11, 5.12, 5.13 and 5.14). The histological study of pancreatic sections of glibenclamide treated group resembled extract treated groups in shape, size, and texture (Fig. 5.15).

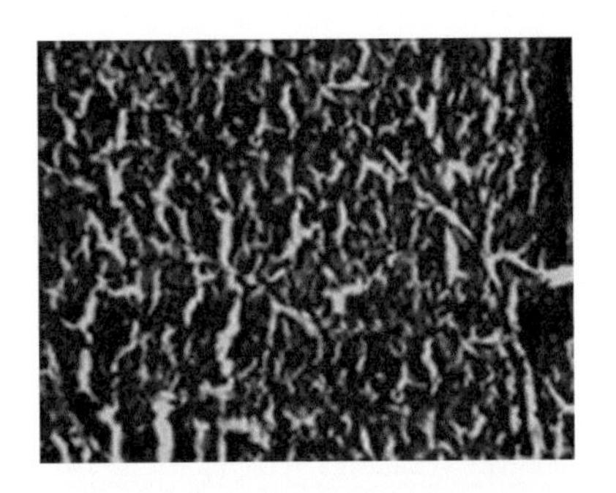

Fig. 5.9 Normal acini and normal cellular population into the islets of Langerhans in pancreas of vehicle-treated rats

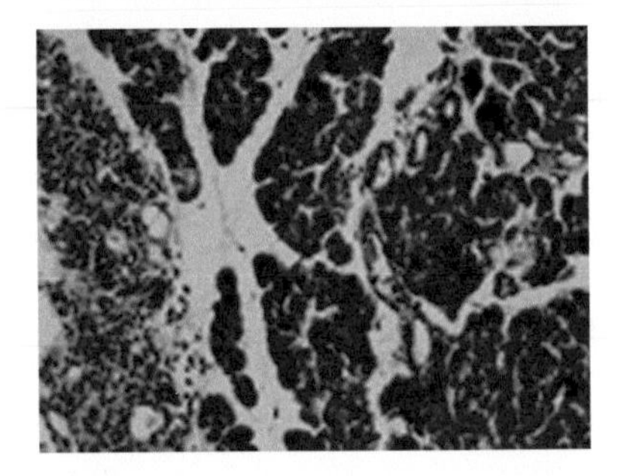

Fig. 5.10 Extensive damage to the islets of Langerhans and reduced dimensions of islets

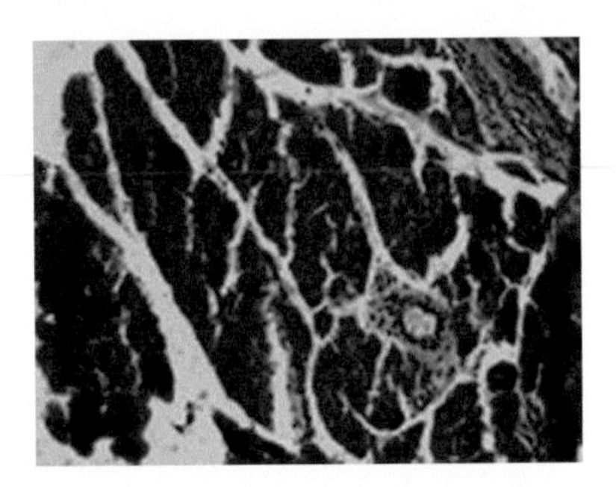

Fig. 5.11 The partial restoration of normal cellular population and enlarged size of β-cells with hyperplasia are shown by hydroethanolic extract (250 mg/kg)

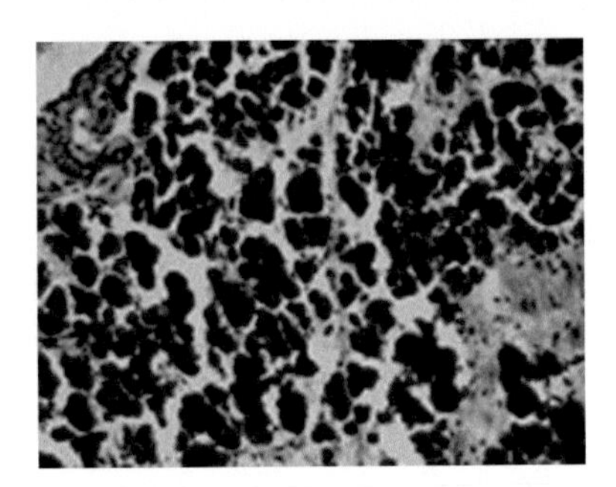

Fig. 5.12 The partial restoration of normal cellular population and enlarged size of β-cells with hyperplasia are shown by hydroethanolic extract (500 mg/kg)

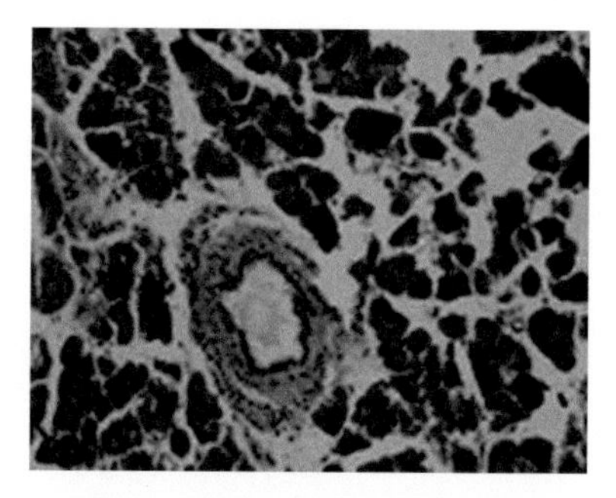

Fig. 5.13 Methanolic extracts (250 mg/kg)

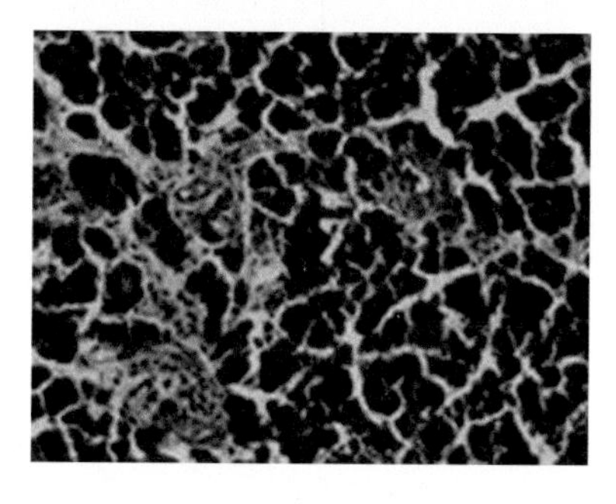

Fig. 5.14 Methanolic extracts (500 mg/kg)

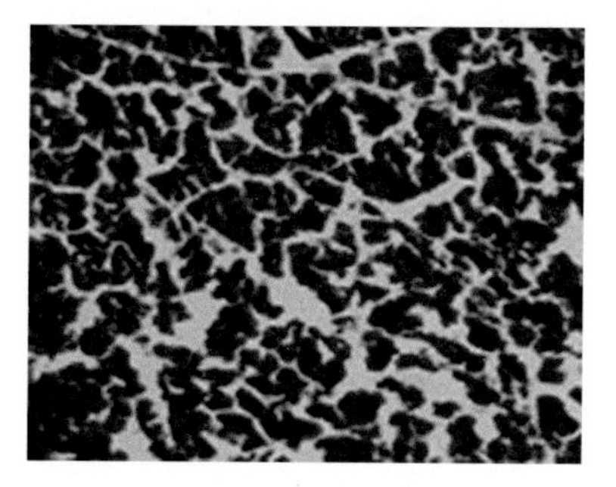

Fig. 5.15 Restoration of normal cellular population size of islets with hyperplasia by glibenclamide

5.2.11 Oral Glucose Tolerance Test (OGTT)

Table 5.11 shows the OGTT study; blood glucose concentration in all groups reached peak levels after 30 min of glucose administration (2 g/kg) and then began to decrease. As compared to normal group, the glucose levels of experimental rats treated with extracts and glibenclamide showed a steep reduction. Hydroethanolic extract (500 mg/kg of b.w.) showed more significant antihyperglycemic activity than other extracts and glibenclamide treated rats. Hydroethanolic and methanolic extracts were found to lower the glucose levels in a dose dependent pattern (Table 5.12).

Table 5.11 Effect of extract of *Artemisia amygdalina* and glibenclamide on glucose tolerance of rats

Groups	Blood glucose level (mg/dL)					
	0 min	30 min	60 min	90 min	120 min	Variation
Normal	82.35 ± 3.44	147.58 ± 4.11	133.60 ± 5.05	125.34 ± 6.40	118.36 ± 3.53	43.72
Pet ether extract (500 mg/kg b.w.)	83.14 ± 2.82	138.58 ± 6.84	127.13 ± 2.46	118.93 ± 5.72	90.57 ± 3.44	8.93
Ethyl acetate extract (500 mg/kg b.w.)	80.76 ± 2.10	142.48 ± 3.53	125.71 ± 6.40	112.91 ± 5.90	94.64 ± 6.35	17.18
Hydroethanolic extract (500 mg/kg b.w.)	83.245 ± 3.71	110.68 ± 3.00**	97.79 ± 3.49**	89.51 ± 4.20**	79.51 ± 4.52**	−4.48
Methanolic extract (500 mg/kg b.w.)	80.47 ± 3.22	122.27 ± 4.20	112.72 ± 4.38	98.61 ± 5.99	86.44 ± 4.65**	7.414
Glibenclamide (3 mg/kg b.w.)	81.89 ± 3.00	118.82 ± 3.85**	107.82 ± 3.67**	98.24 ± 4.29**	82.56 ± 4.11**	0.81

The values are expressed as mean ± SEM; $n = 5$ in each group
$^{**}P < 0.01$ as compared with normal at the same time (one-way ANOVA followed by Dunnett's multiple comparison test)

Table 5.12 Dose-dependent effect of hydroethanolic and methanolic extracts of *Artemisia amygdalina* on glucose tolerance of rats

Groups	Blood glucose level (mg/dL)					
	0 min	30 min	60 min	90 min	120 min	Variation
Normal	81.25 ± 2.97	149.49 ± 3.24	136.64 ± 4.22	128.44 ± 5.53	115.42 ± 6.30	42.05
Hydroethanolic extract (50 mg/kg b.w.)	82.41 ± 3.23	145.45 ± 4.67	131.24 ± 2.35	122.46 ± 4.86	112.25 ± 4.33	36.20
Hydroethanolic extract (100 mg/kg b.w.)	83.67 ± 4.15	132.56 ± 4.32	120.18 ± 4.50	110.16 ± 4.34	101.14 ± 5.36	6.88
Hydroethanolic extract (250 mg/kg b.w.)	81.54 ± 5.25	121.24 ± 2.96**	107.73 ± 4.39**	96.64 ± 3.74**	87.15 ± 2.54**	6.88
Hydroethanolic extract (500 mg/kg b.w.)	82.42 ± 4.64	108.68 ± 3.38**	96.21 ± 3.44**	83.31 ± 4.72**	72.15 ± 5.42**	221212.46
Methanolic extract (50 mg/kg b.w.)	83.66 ± 4.12	150.52 ± 3.35	140.88 ± 3.48	128.25 ± 2.78	118.26 ± 3.35	41.35
Methanolic extract (100 mg/kg b.w.)	82.52 ± 4.45	141.65 ± 3.22	129.42 ± 3.68	116.27 ± 3.85	105.51 ± 5.22	27.86
Methanolic extract (250 mg/kg b.w.)	83.24 ± 5.24	131.57 ± 4.38	116.21 ± 5.12	106.35 ± 4.62	95.23 ± 4.95**	14.40
Methanolic extract (500 mg/kg b.w.)	82.57 ± 6.35	120.09 ± 5.14	103.41 ± 2.56	92.35 ± 3.44	81.13 ± 2.55**	1.74
Glibenclamide (3 mg/kg b.w.)	80.68 ± 2.75	116.78 ± 5.83**	104.23 ± 6.73**	94.16 ± 5.25**	83.25 ± 4.75**	3.18

The values are expressed as mean ± SEM; $n = 5$ in each group
**$P < 0.01$ as compared with normal at the same time (one-way ANOVA followed by Dunnett's multiple comparison test)

5.3 Conclusion

In conclusion, *A. amygdalina* plants both in vitro grown, as well as those acclimatized in the greenhouse reveals antioxidant activity against hydroxyl, superoxide, and lipid peroxyl radicals. Further, the plants showed anti-inflammatory and also immunosuppressive activity. Based on the above results and analysis, it can be concluded that *A. amygdalina* has the potential to suppress cell-mediated immunity as well as humoral immunity, and it may be a potential therapeutic candidate in several immune stimulant clinical conditions. From the current study, it can be said that this plant may be a good resource of bioactive components especially flavonoids which have been found to have the immunomodulatory and anti-inflammatory activity. Further, the common symptoms of diabetes, that is, polyphagia, polydipsia, and weight loss, have been found to be lessened by the extracts of *A. amygdalina* (hydroethanolic and methanolic extracts each at a dose level of 500 mg/kg of b.w.) in diabetic rats. The extracts significantly reduced fasting glucose levels in diabetic rats and also reduced the lipid profile parameters in diabetic rats. The extracts were found significantly decreasing the activities of SGPT, SGOT, and ALP in diabetic rats. In conclusion, the histopathological investigation along with the biochemical evaluations suggests the strong anti-diabetic potential of *A. amygdalina*. The results observed show the effect on both the pancreatic-β-cells and the blood glucose level.

References

Bhagwat DP, Kharya MD, Banietal S (2010) Indian J Pharmacol 42:21–26

Gazafar K, Ganaie BA, Seema A, Mubashir K, Showkat AD, Younis Dar M, Tantry M (2014) BioMed Res Int 1–10

Jayathirtha MG, Mishra SH (2004) Phytomedicine 11:361–365

Lunardelli A, Leite CE, Pires MGS, deOliveira JR (2006) Inflamm Res 55:129–135

Mangathayaru K, Umadevi M, Reddy CU (2009) J Ethnopharmacol 123:181–184

Mubashir K, Ganaie BA, Gazafar K, Seema A, Akhter HM, Akbar M (2013) ISRN Inflammation 2013:1–6

Organisation for Economic Cooperation and Development (2008) OECD guidelines for the testing of chemicals Test 425

Rasool R, Ganaie BA, Akbar S, Kamili AN (2013) Chin J Nat Med 11:0377–0384

Suffredini AF, Fantuzzi G, Badolato R, Oppenheim JJ, O'Grady NP (1999) J Clin Immunol 19:203–214

Thakur M, Connellan P, Deseo MA, Morris C, Dixit VK (2011) Evidence-based complementary and alternative medicine 7:1–7

Winter CA, Risley EA, Nuss GW (1962) Proc Soc Exp Biol Med 111:544–547

Printed in the United States
By Bookmasters